Snowmen Are Cool: Woodcarving Winter Creations

Paul and Camille Bolinger

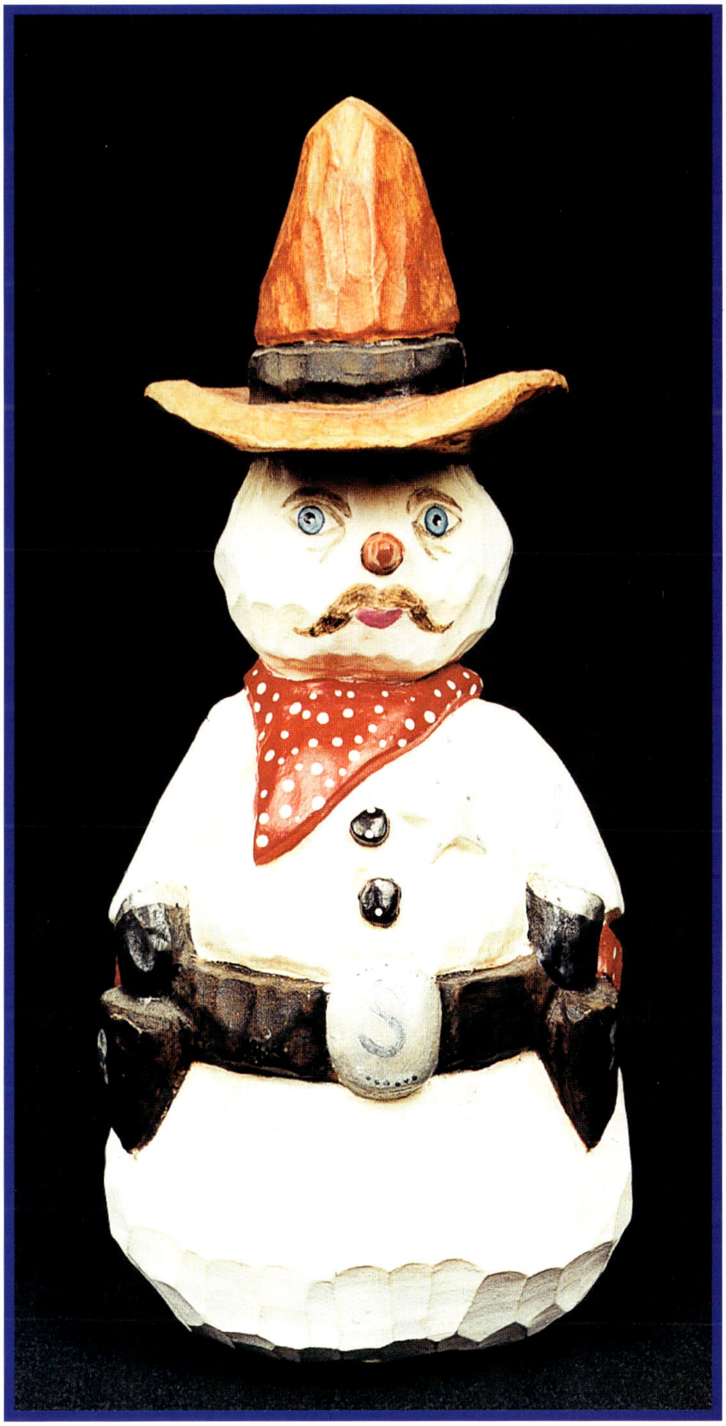

Schiffer Publishing Ltd
4880 Lower Valley Road, Atglen, PA 19310 USA

Acknowledgements

Snow was covering the ground when Pat McChesney arrived in his boots, carrying all his camera equipment, ready to photograph yet one more book. Pat has photographed every one of our five previous books and, unbelievably, he wanted to do this one too. I enjoy Pat's company so it is a great pleasure when he comes over and hangs around for the three days it takes to get all the pictures for a book. He turns work into fun. Thanks Pat.

Dedication

Recently my wife Camille and I were discussing all the things I had listed for her to do while I was away on a trip. All of a sudden something I have always known became even more obvious to me - I couldn't do it without her. She does everything on her own list and then finds time to help me by painting the pieces we create; amazingly she is good at it all too. Thanks Camille for everything.

Contents

Library of Congress Catalog Number: 99-64923

ISBN: 0-7643-0976-5
Printed in China

Typeset in Korinna BT \Humanist 521 BT
Designed by "Sue"

Published by Schiffer Publishing Ltd.
4880 Lower Valley Road
Atglen, PA 19310
Phone: (610) 593-1777; Fax: (610) 593-2002
e-mail: schifferbk@aol.com
Please visit our website catalog at
www.schifferbooks.com
or write for a free printed catalog.
This book may be purchased from the publisher.
Please include $3.95 for shipping.

In Europe, Schiffer books are distributed by
Bushwood Books
6 Marksbury Avenue
Kew Gardens
Surrey TW9 4JF England
Phone: 44 (0)181 392-8585; Fax: 44 (0)181 392-9876
e-mail: bushwd@aol.com

Please try your bookstore first.

We are interested in hearing from authors
with book ideas on related subjects.

Chapter 1

So Round, So Firm, So Fully Packed

In the past few years I have added carved Snowmen to our line of resin-cast reproduction gift items in response to the growing popularity of Snowmen and the need created by Snowmen collectors; yes there are people who specialize in collecting Snowmen. Part of the fun of collecting and displaying Snowmen comes from the fact that they are not tied to the Christmas season and, therefore, do not need to be put away after the season ends. Snowmen can be kept out and displayed until the end of winter.

The Snowmen in the collection pictured in this book are generally humorous. The humor gives collectors an additional reason to buy since there must be some natural limit to how many generic Snowpeople one can own. The possible variations in humor for Snowmen are almost endless. Take a look at my designs and let your mind wander.

Note: Please understand that many of these designs have been reproduced and being sold around the country. All of the designs are copyrighted. This book is intended to show you how to carve your own snowmen. You are welcome to use the designs for study, practice, and your own private purposes but please do not make copies of my designs for resale or commercial purposes.

So what is it about real Snowmen, the cold ones you build outside, that endears them to us so much we want to make little wooden replicas of them? I guess one major factor about the rising popularity of Snowmen is their universality. Anyone who has ever lived or traveled where the snow falls has seen or, perhaps, even built a snowman. What great fun it is to start with a small snowball and roll it around until it gets big enough that you have to stop. Of course you have to roll a couple more snowballs and somehow get them perched in the right place to make the Snowman.

There is joy in creating something from nothing. There is a similarity between making Snowmen from the nothingness of snow and carving them from a formless block of wood. The great thing about the wooden ones is that they will last forever – unless you give them away or sell them.

Beyond the universality and joy of creation there are clothing and decorating aspects that can be the most fun. Finding just the right old things to dress a real Snowman or Snowlady in can be great fun. Hats, gloves, mittens, boots, or whatever can be added to your creation to give it color and flair. Hats do not have to be top hats either since just about any kind of hat works as long as it is funky enough to make a statement. The same fun comes when designing your snow carving.

All the fun of making the real Snowmen can be translated into making the carved Snowmen shown in this book. I carve my Snowmen from Basswood since that is my favorite carving wood, however, the type of wood isn't critical. Having fun is the critical element in carving Snowmen; here is a chance to let it all hang out and just do pretty much what ever comes to mind.

The form of a Snowman can be anything from the traditional stacked balls to a simple blob as long as it looks pleasing. I have made different shaped Snowmen but I prefer the shaped bodies you see in the Igloo Snowman and the Snowbank Snowman, which are the focus of this book. You can try any or many of these shapes and experiment with the shapes until you find one that pleases you.

One of the important decisions you need to make when starting to design your Snowman is whether you will use the "snow" arms that I have used in these carvings or whether you will use "twig" arms. Twig arms have a great look to them but in some ways they limit what you can do with the carving. With the "snow" arms you can use them to hold and display things that will add to the uniqueness of your carvings. The choice is yours unless you want to make so many Snowmen that you can make some of each type. I used the "snow" arms because they are easier to reproduce in casting and less easily broken in transporting and selling.

Another decision is what kind of face you are going to put on your Snowman. The traditional face made from blobs of coal work well and are simple to make. More elaborate faces are more difficult but can add interest to your Snowman. I try to get Camille to paint more elaborate and cuter faces because they attract the buyer's interest more. You may want to experiment until you find a look you like. Try drawing various faces before you carve the head and especially draw some different styles of eyes.

From faces to feet; another decision is what to do with the bottom of the Snowman. Does he just end as a blob on the ground or does he have boots, shoes, or something else? In one of my examples you see a boot sticking out while the other foot is hidden behind an igloo. You should decide what you are going to do before you start. The simplest approach is just to leave the bottom as a blob on the first carving you do then you can graduate to more elaborate bottoms later. Details at the bottom do add interest and make the Snowman seem more "alive." You might even think about adding skis, snowshoes, skates, a sled, or a toboggan for interest.

Chapter 2

Techniques and Tips

I used the "saw out and glue up" technique for the snowmen in this book. This is the same technique I used in our book "The Edge of Christmas." I begin this technique by creating a complete drawing of the piece, then I "decompose" or separate the drawing into layers that can be sawn from 3/4" basswood stock. The cut out 3/4" layers are then glued up to form the blank for the carving.

You will find it possible, of course, to carve the Snowmen directly from a thicker block; it just means you have more work to do. I do this in some cases, however, I find that the carving part goes faster and I seem to get a better product when I start from a glued up blank like the ones shown in this book. There is preparation time in the cutting and gluing, of course, but that goes pretty fast and is offset by the time saved in carving.

I always place the base of the drawing, for a layer, along a straight line marked perpendicular to the grain of the 3/4" stock. This gives me a flat, true edge on every layer that I can use as the baseline for gluing up the layers. I use plain Elmer's wood glue and clamp the pieces, generally leaving them clamped overnight. The glue carves away easily and any glue lines are covered by the undercoat and paint that Camille applies.

Take a look at the patterns for the "Igloo Santa." The blank is made up of three layers. The first layer is the full front and includes the outline for the sign. Layer two is almost the same except the sign is not needed. Some "arm" is needed behind the sign. The third layer has the least wood because the boot and the igloo are not needed and neither is the sign.

In the patterns for the "Snowbank Snowman" you see that very little exists on layer one. In fact, the real reason for the layer is to give some depth to the carving by allowing the scarf, the sign, and the boot to stick out in the front. Layer two is the one with the face drawn on it and has the medium high snowbank; note that very little snowbank is to the left of the figure. The last layer has the full outline of snowbank on both sides of the Snowman.

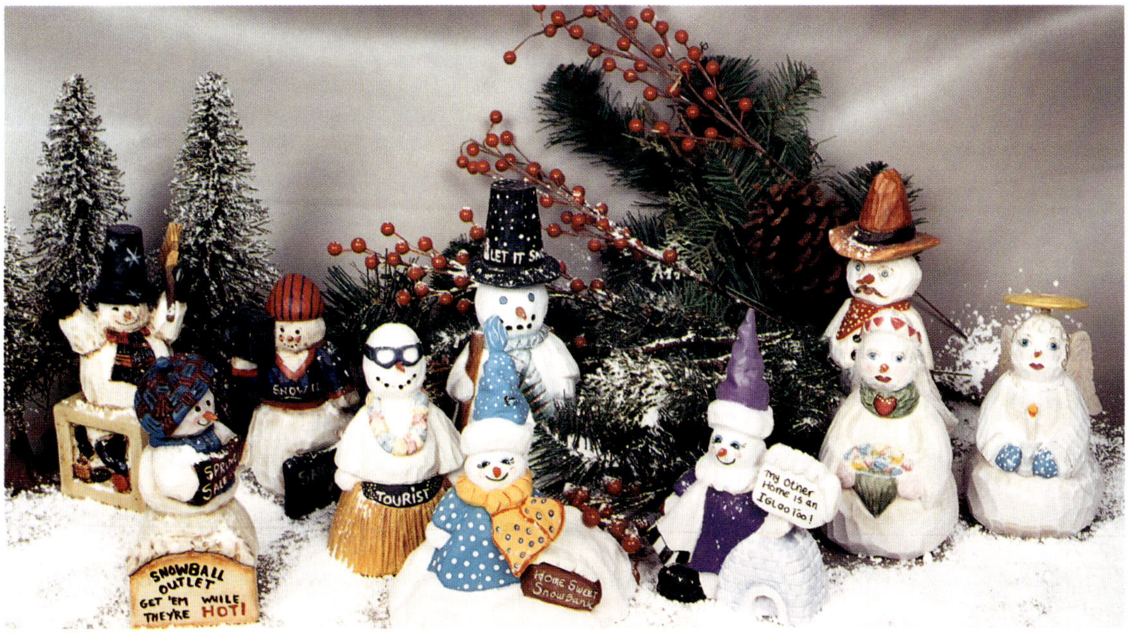

A group photo of our snow friends.

Chapter 3

Carving the Igloo Snowman

These photos show the step-by-step carving I did for the Igloo Snowman. I use a carver's arm and a carver' s screw to hold the piece while I carve. This technique has served me well for many years and is an inexpensive way to hold the work. I use carver's gouges along with a mallet to get the most carving for the least wear and tear on me. I carve many hours a week and need to be careful about fatigue, carpal tunnel, and other threats. Occasionally I use a knife and sometimes find myself holding smaller pieces while I carve some of the areas.

The process of carving is one of "peeling the onion;" by that I mean going layer by layer in a systematic fashion. I work around the piece without concentrating on any particular area. This way I can keep the piece in balance without taking too much off of one area.

Follow the instructions and examine the pictures closely. Good luck.

Here is the Igloo Snowman ready to carve. I drew the lines in heavy ink after I cut him out so they would be easier to see. He is sitting on top of my carver's arm that is secured to the workbench with clamps. I find the carver's arm a perfect method of holding the work except in cases where the piece is very small.

The carver's arm shown here is a commercial version but you can make one easily by cutting a slot into a 4x4 piece of good hardwood like oak.

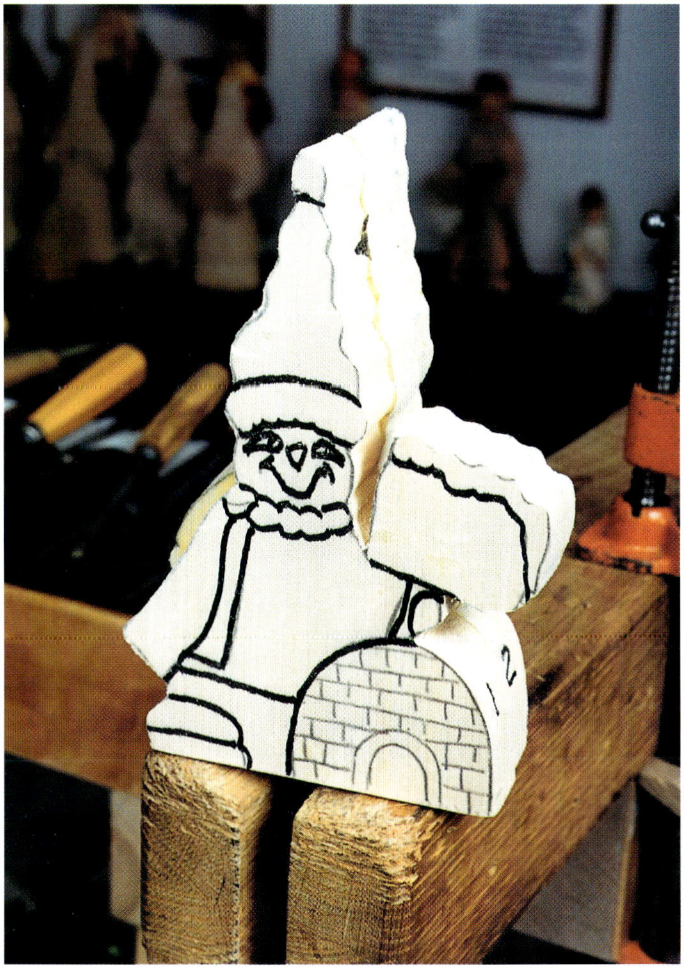

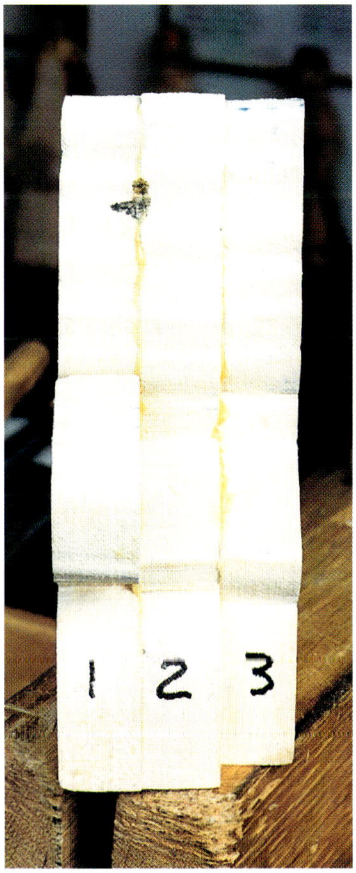

This is the side view of the Igloo Snowman showing the three layers that were glued together to make the blank.

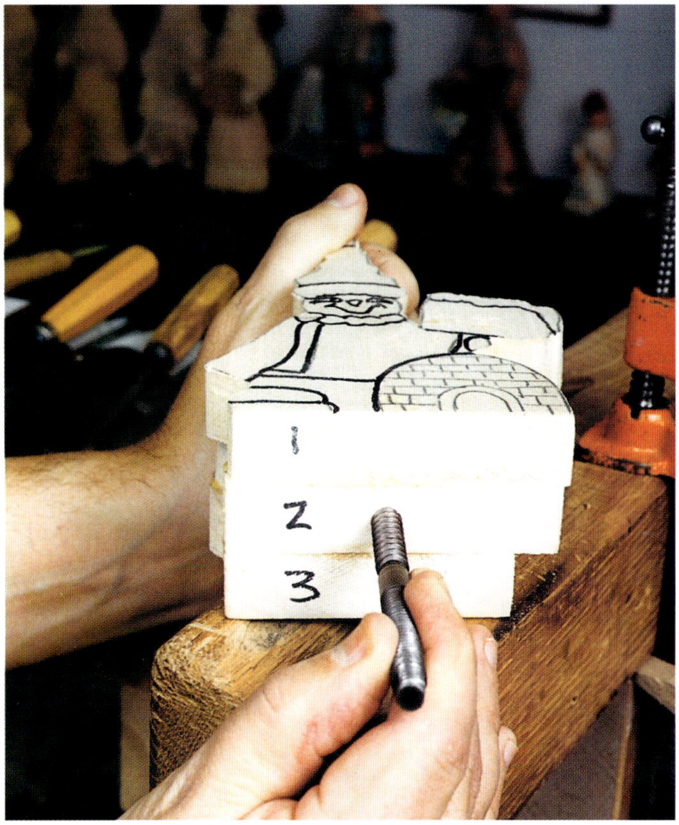

Here is the bottom view of the Igloo Snowman showing how the carver's screw has been inserted into the bottom of the piece. This particular screw is a 3/8" carver's screw. The carver's screw is half wood screw and half machine screw so that the wood screw portion can go into the piece while the machine screw portion is used to attach the handle for tightening. I drilled a 1/4" pilot hole for the carver's screw.

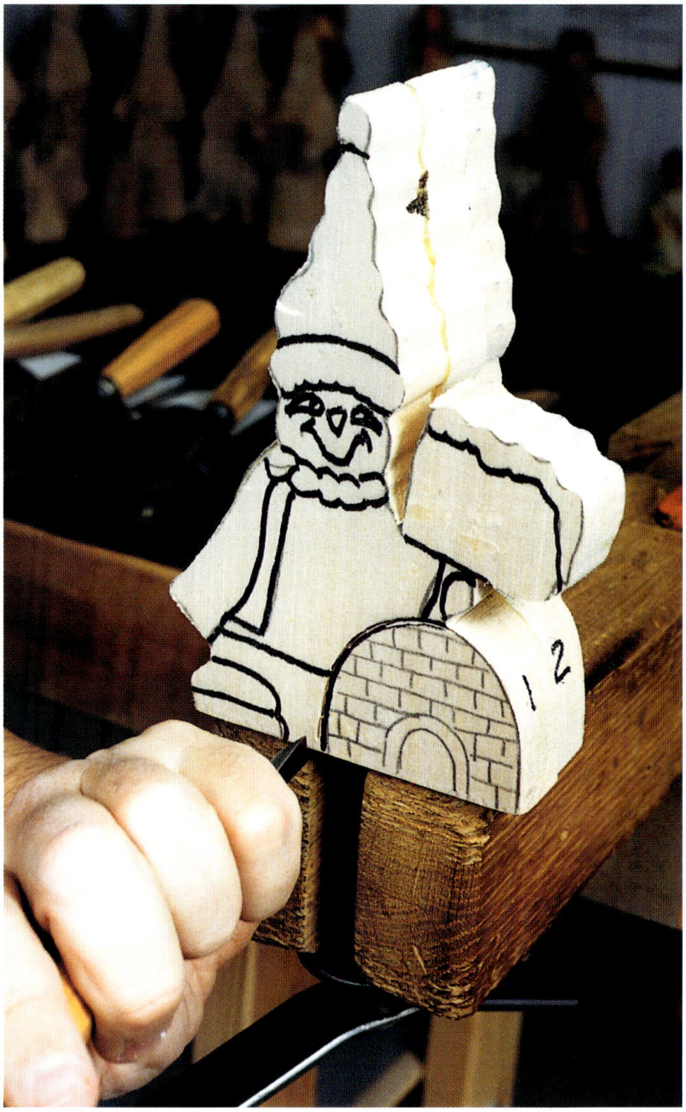

Here I start by outlining the Igloo with a "U" gouge. Note that the handle of the carver's screw has been attached and tightened up against the bottom of the carver's arm. I use a very large washer to keep the handle from crushing the wood on the bottom of the carver's arm.

Carving is generally a process of establishing a "stop cut" then cutting the wood away until the "stop cut" is reached. The purpose of the "stop cut" is to keep the cut from "running" up the wood thus removing too much wood.

To proceed on a carving like this you must have a clear impression in your mind about which areas of the piece will be the farthest "forward" or closest to you. Here the boot tip will clearly be far forward as will be the igloo.

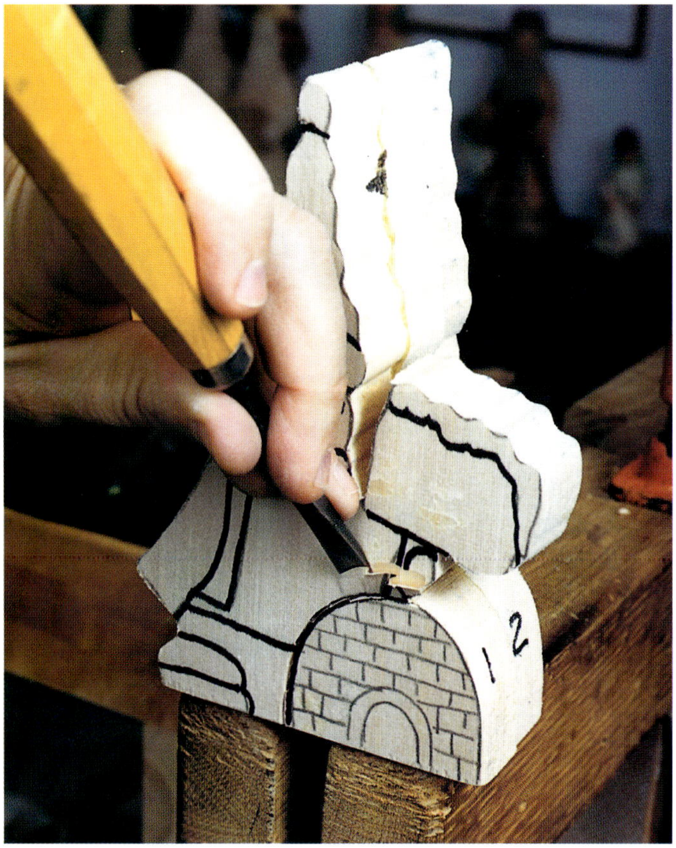

Remove the wood outside the "stop cut" with a "U" gouge.

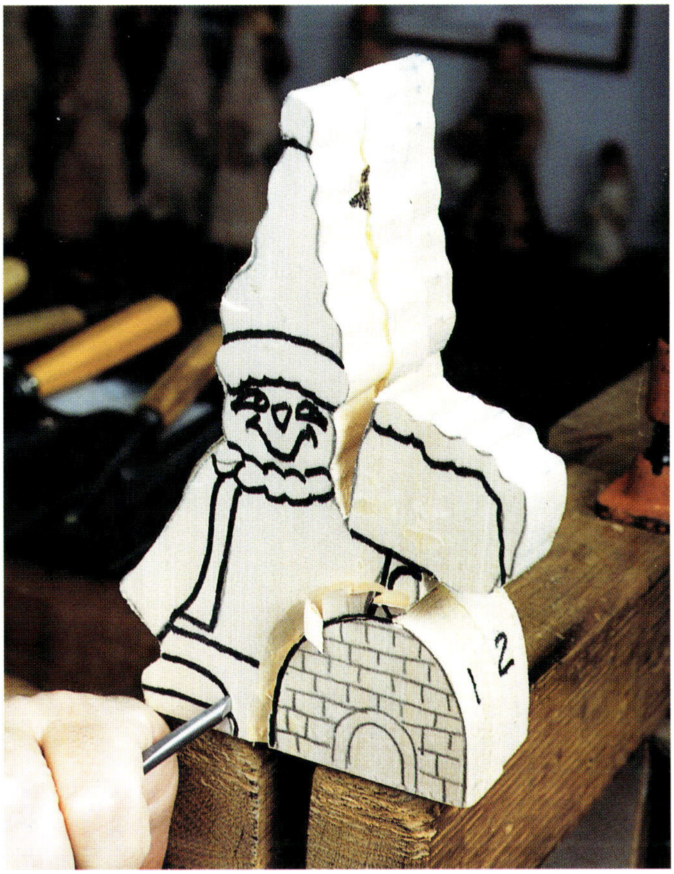

Outline the boot tip with a small "U" gouge. Remove the wood from around the boot tip so that the tip will be left forward.

Outline the edge of the scarf using a knife. Remove the wood from the Snowman's belly with a small "U" gouge so that the scarf stands out. The scarf will be a prominent feature of the carving when painted.

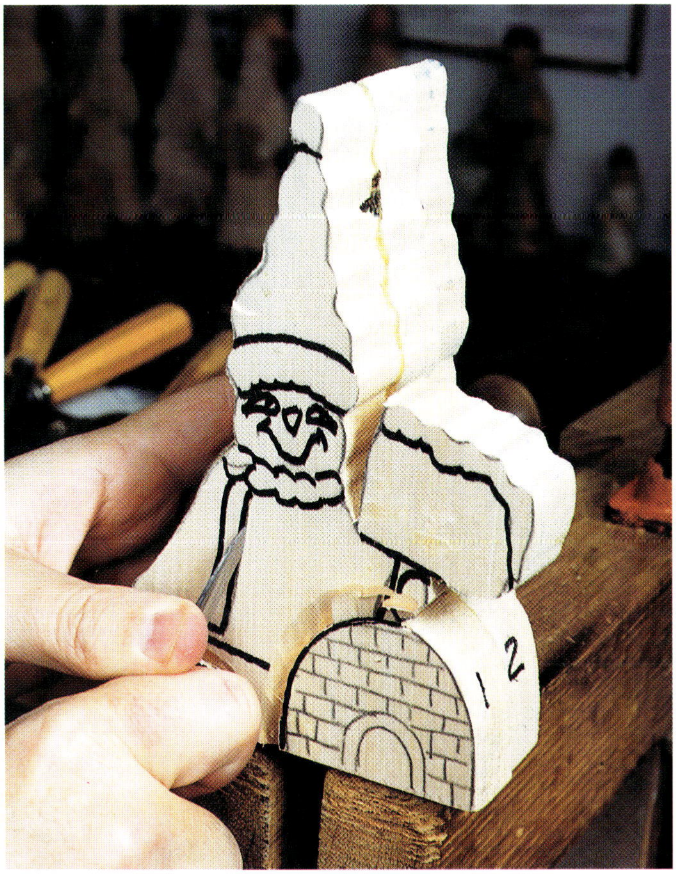

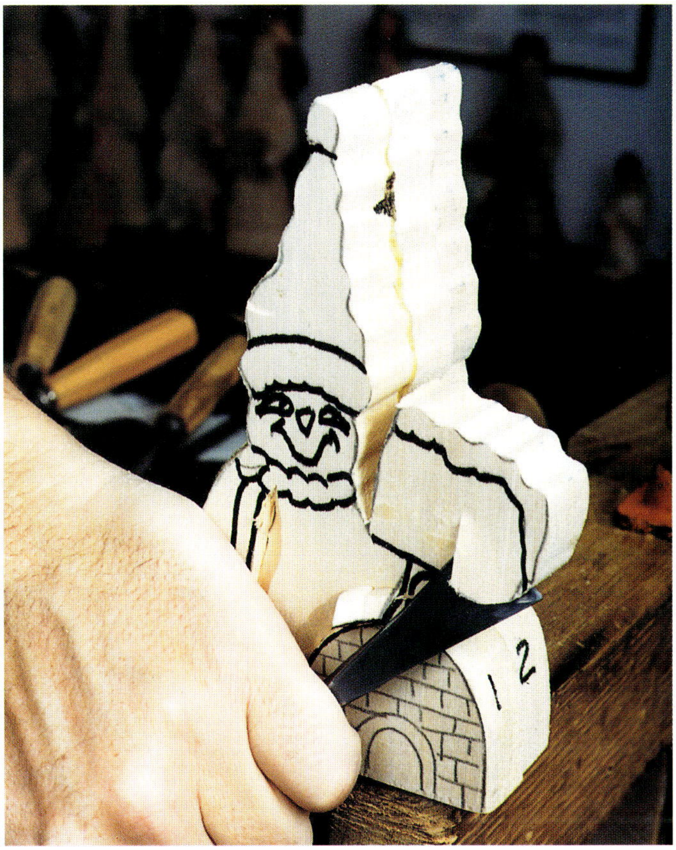

Use a large shallow "U" gouge to cut back the sign. The sign will lean back somewhat so that it is easier to read.

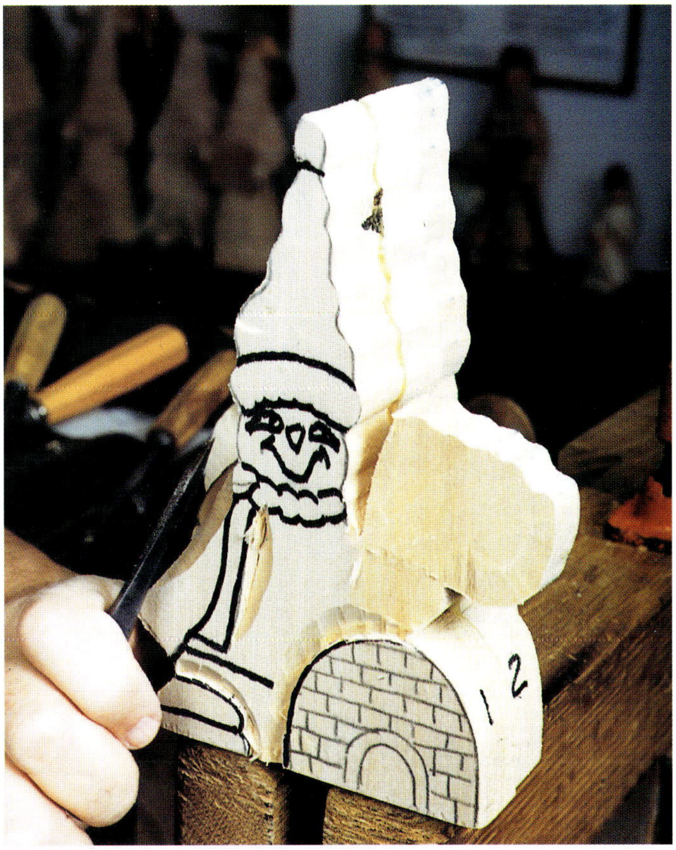

Start cutting back the left side of the scarf. It will lean back to the left when completed.

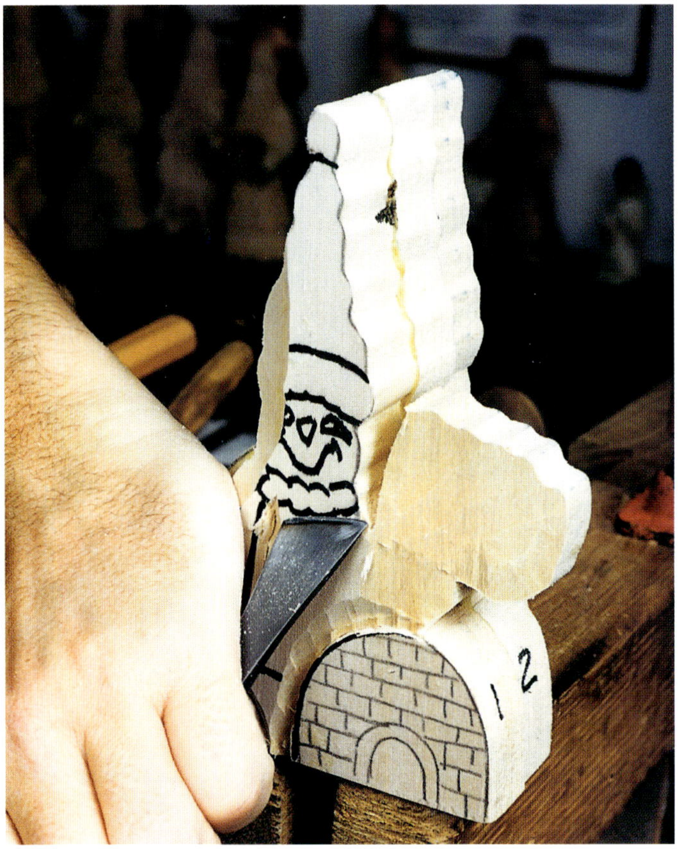

Use a large shallow "U" gouge to lean the upper portion of the carving back. Leaning him back will help make the igloo, his belly, and the boot tip more prominent and will make him easier to see when he is done. You will have to re-draw the pattern onto the piece later.

Continue to lean him back. Looking at the glue lines on the side of the carving you see how I have not touched the lower portion of the carving but have moved the top of the hat back past the first 1/3 of depth.

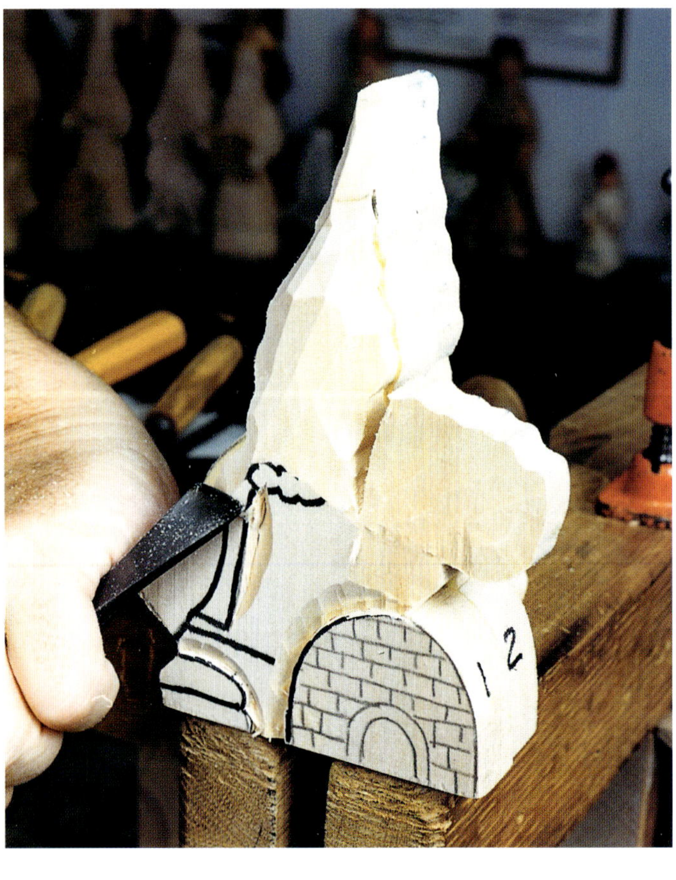

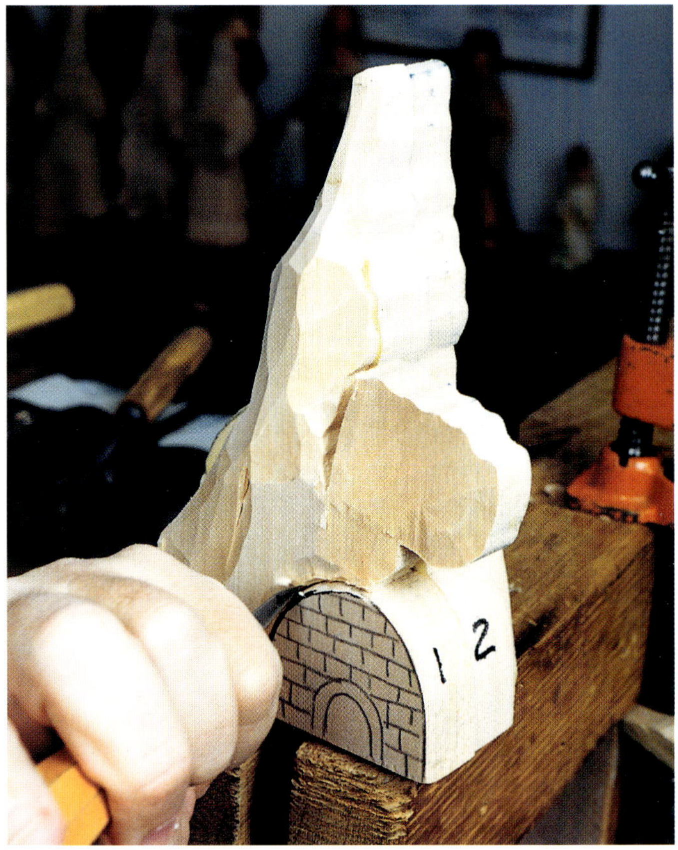

Re-cut the stop cut around the Igloo.

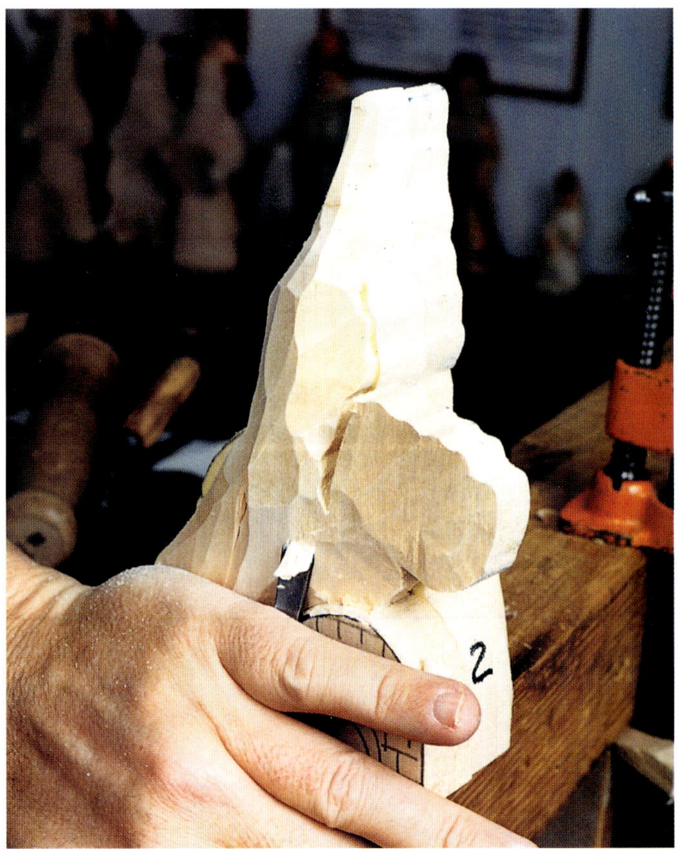

Remove the wood outside the Igloo.

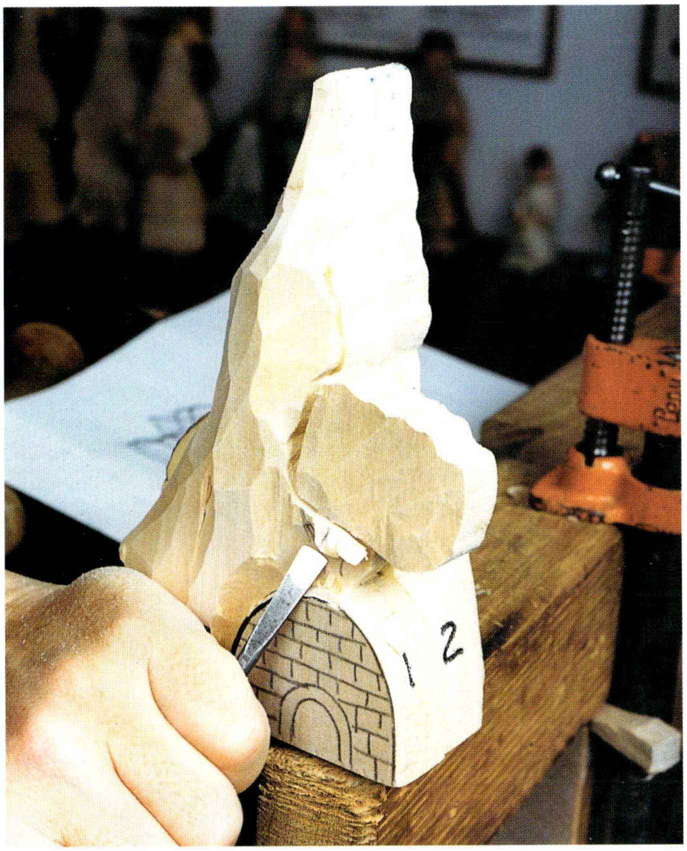

Make a stop cut at the bottom of the sign and then cut up to the stop cut. This will leave the sign face prominently displayed.

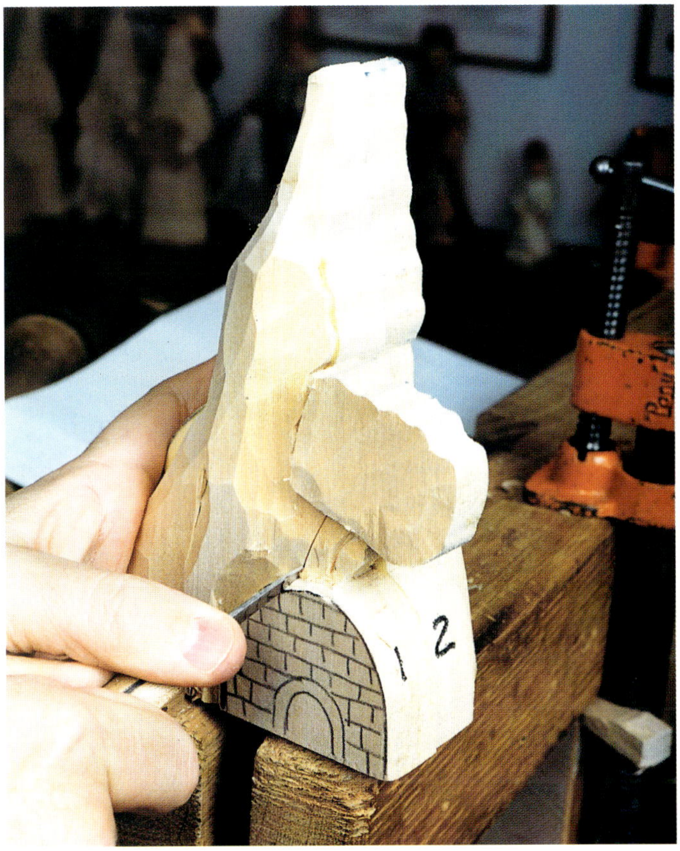

Cut the line at the left side of the stick holding up the sign.

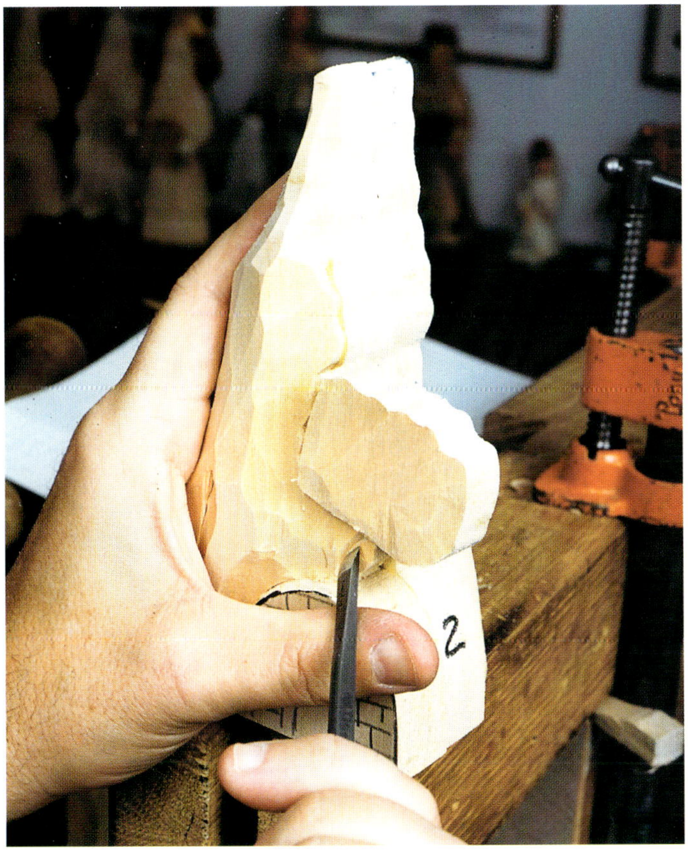

Cut a line separating the Snowman's hand from the stick holding up the sign.

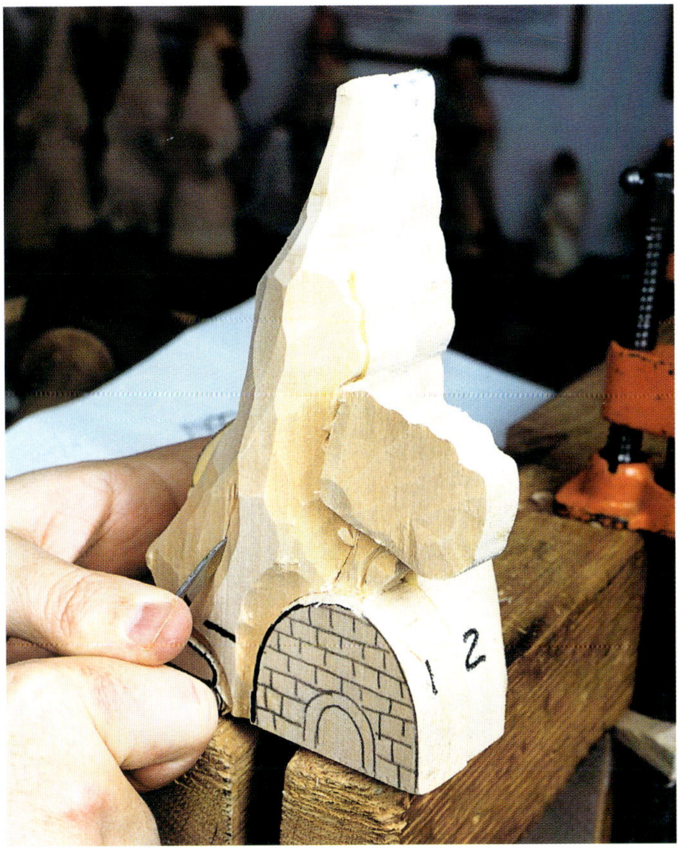

Re-cut the line at the inner edge of the scarf.

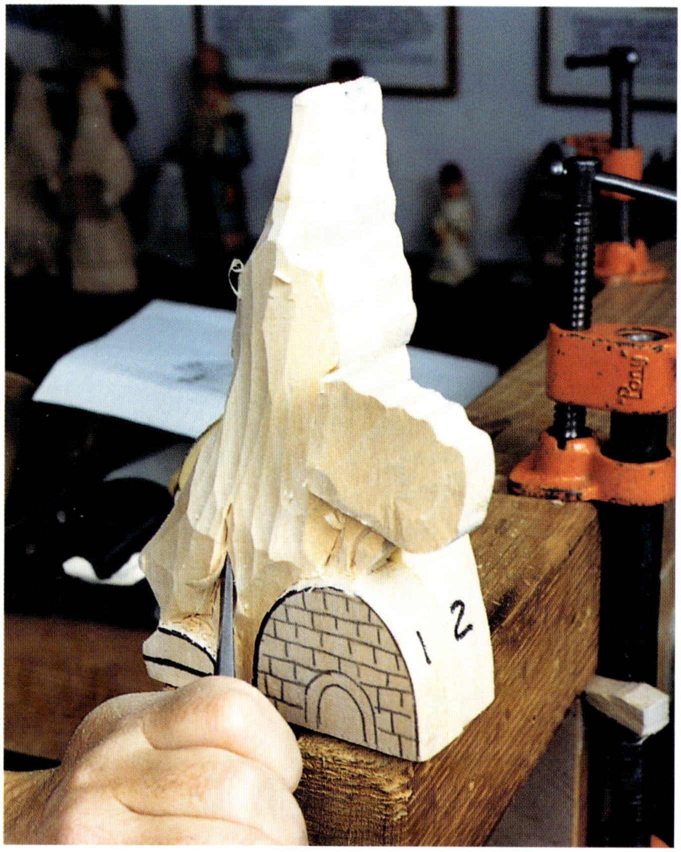

Cut the belly away from the edge of the scarf.

Make a stop cut near the top of the sign that will allow you to separate the sign from some "snow" resting atop the sign. Cut the sign away from below the stop cut to reveal the snow atop it. Here, I re-drew the lines showing the hat, scarf, and bottom of the coat so that I could keep the carving in balance.

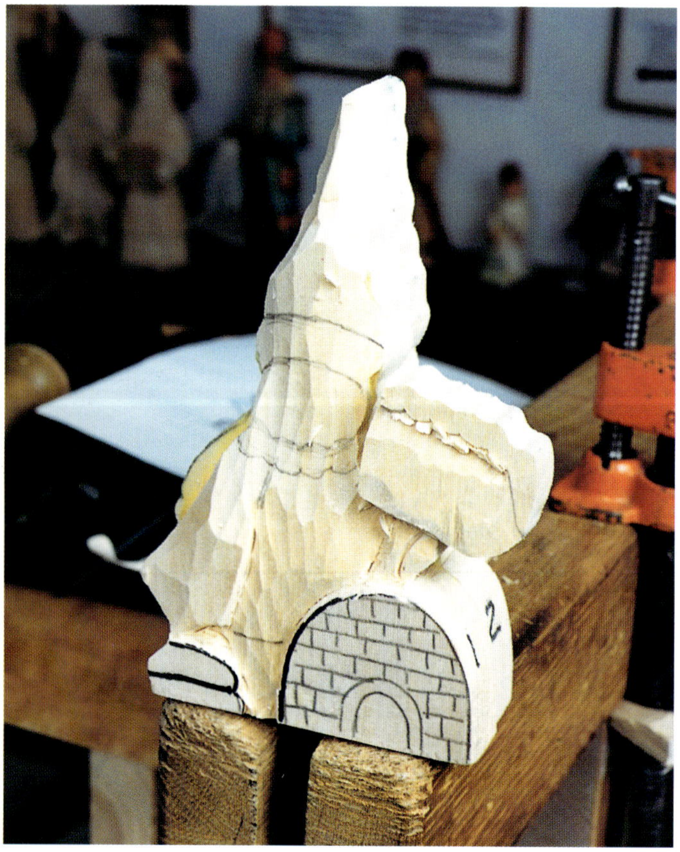

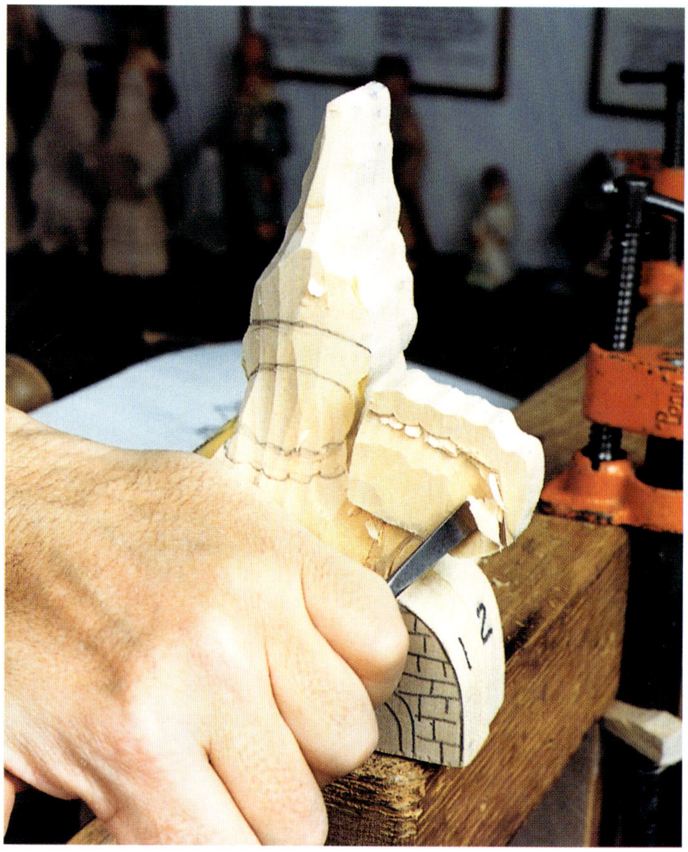

Deepen the line between the sign and the snow.

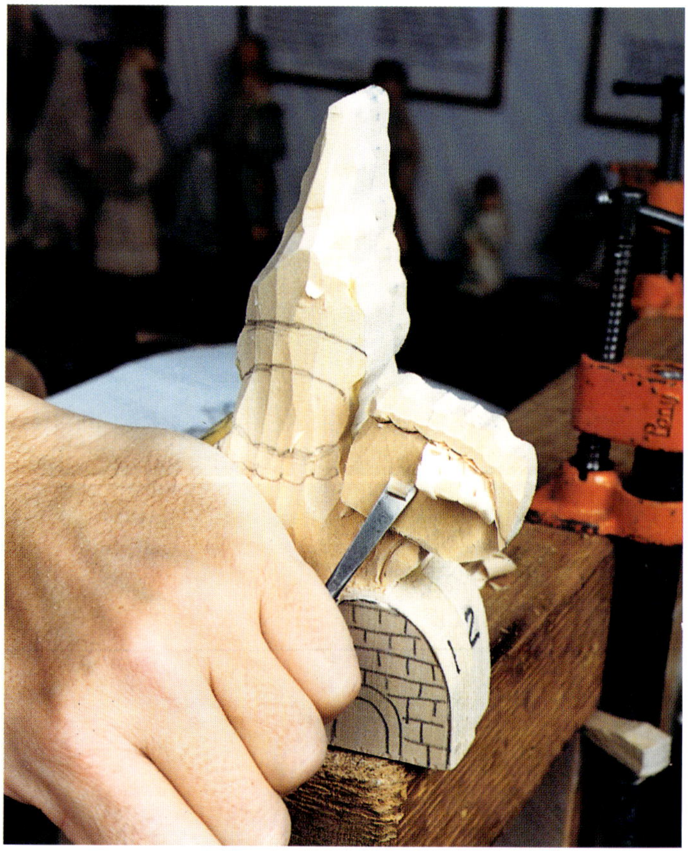

Remove some of the sign face.

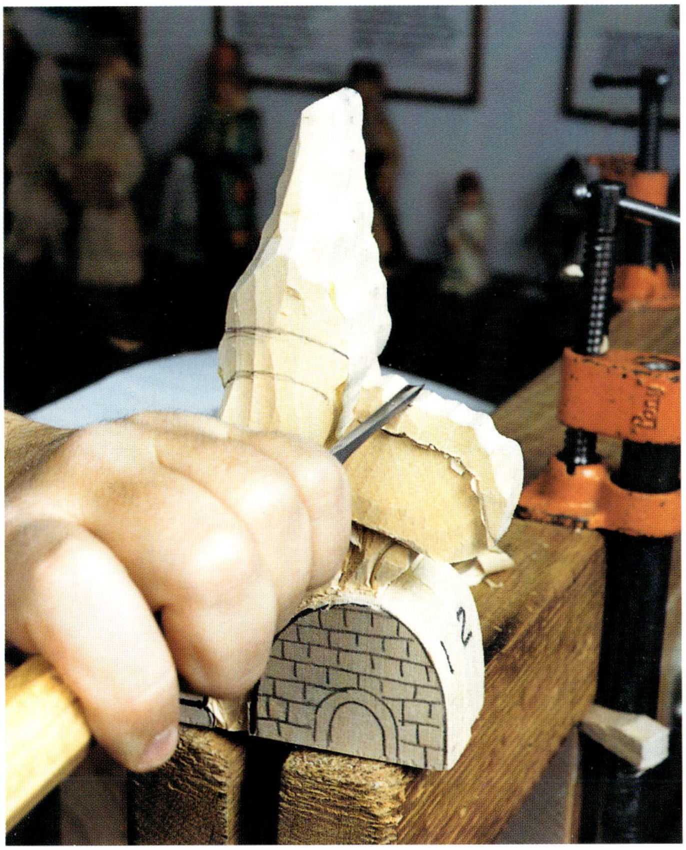

Cut back the top edge of the snow slightly to round it like snow. Use a small "U" gouge.

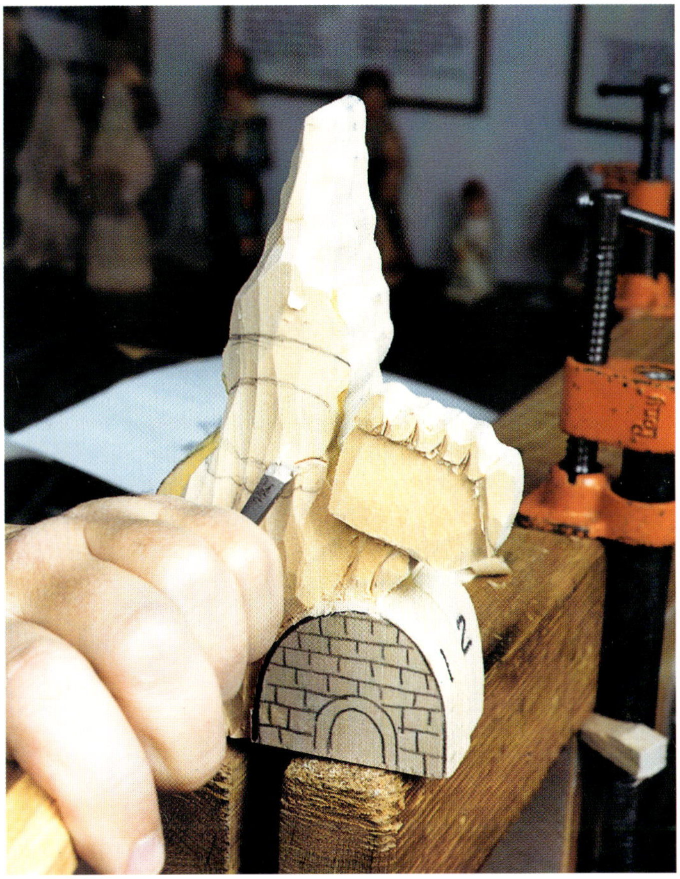

Use a small "U" gouge to outline the scarf around the Snowman's neck.

Cut down the face into the stop cut then cut up the scarf into the stop cut to make the scarf stand out around the neck.

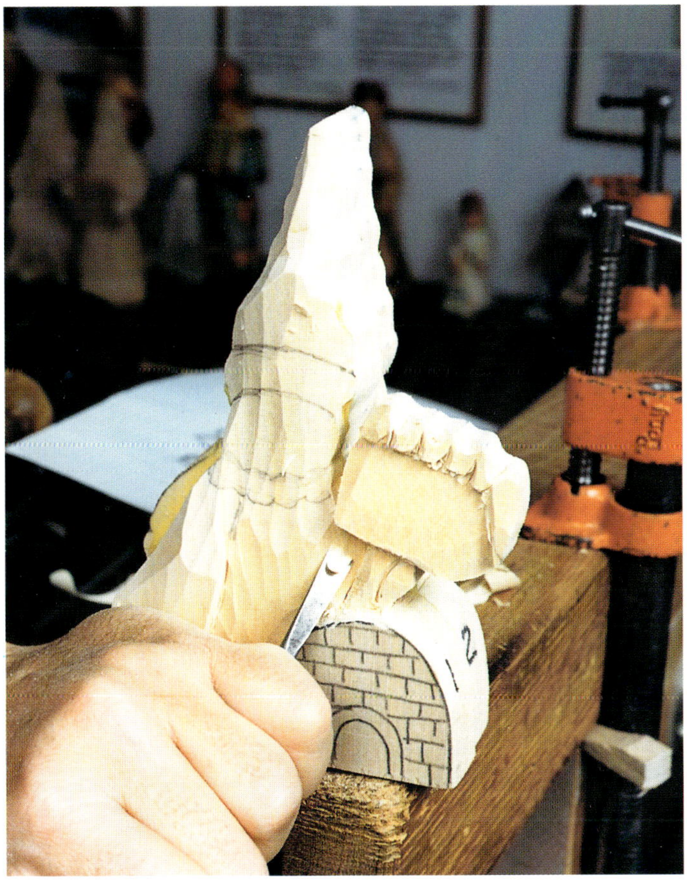

Cut lines into the snow atop the sign to give it more appeal and detail. Deepen the area below the sign and round it to the right. Remember that the snowman is "round, firm, and fully packed."

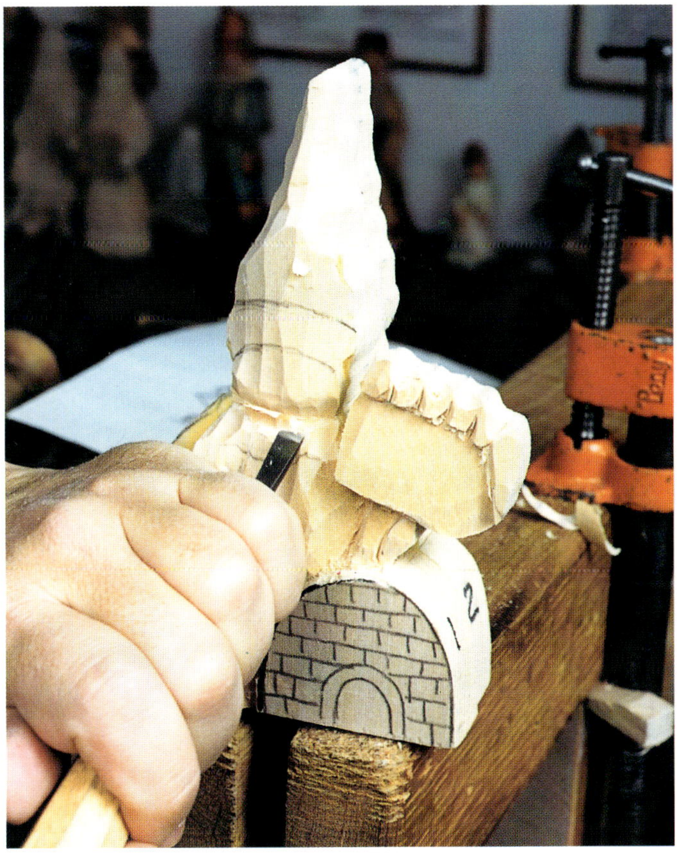

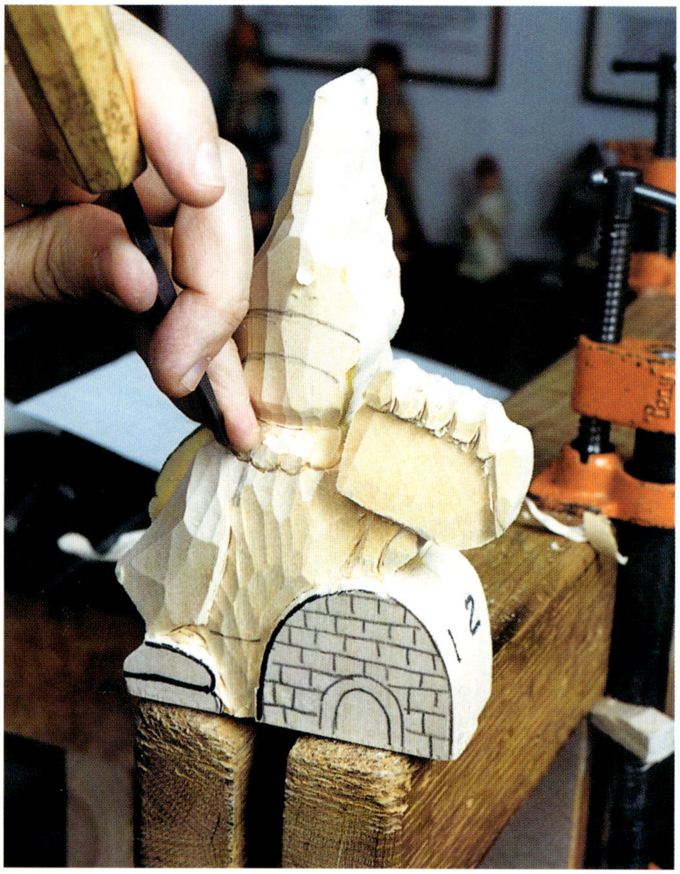

Use a small "U" gouge to outline the bottom of the scarf.

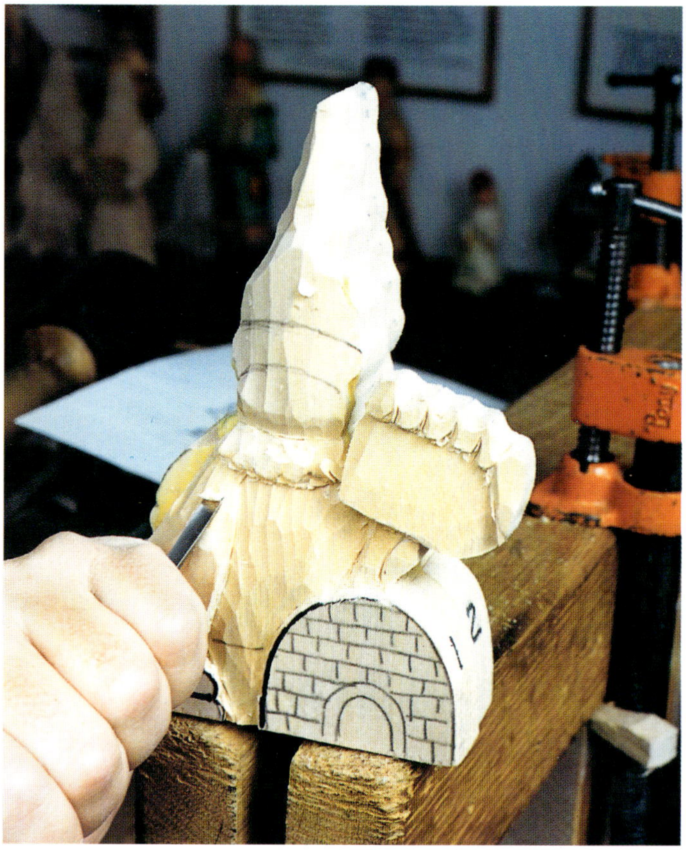

Continue to round the top of the body into the scarf until it looks good to you.

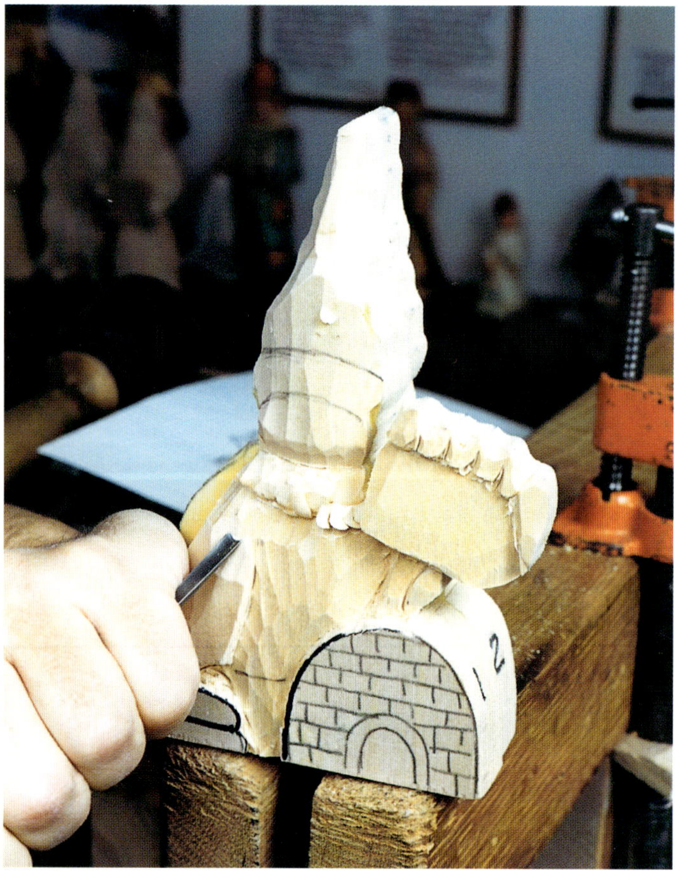

Round the carving up into the bottom of the scarf.

I have moved to the left side of the carving now. From this side view you can see how far back the carving has been cut in order to "lean" – at the top, the whole first 3/4" layer has been removed and the hat actually goes back some into the second 3/4" layer.

Use a large shallow "U" gouge to begin rounding the back corner of the piece.

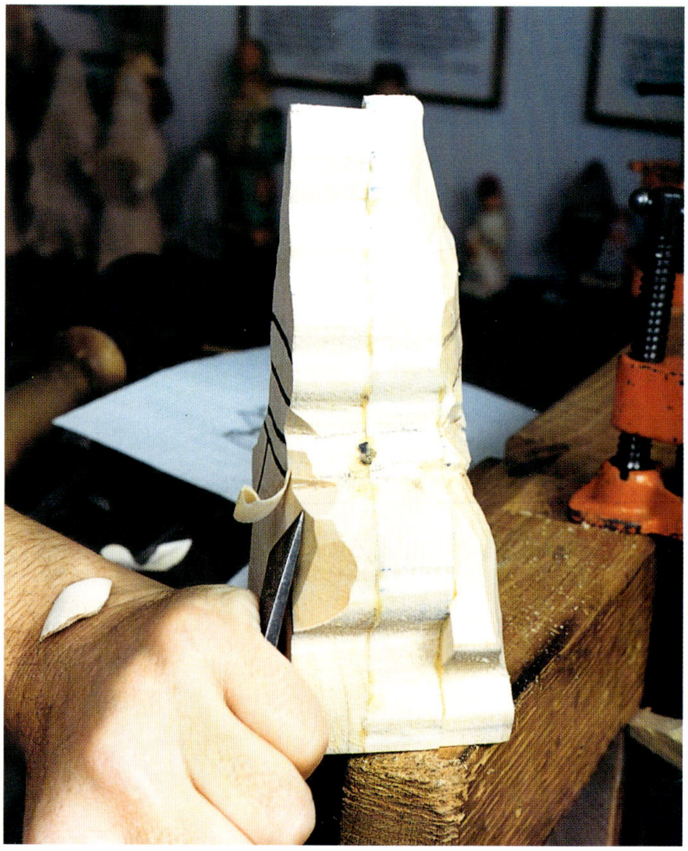

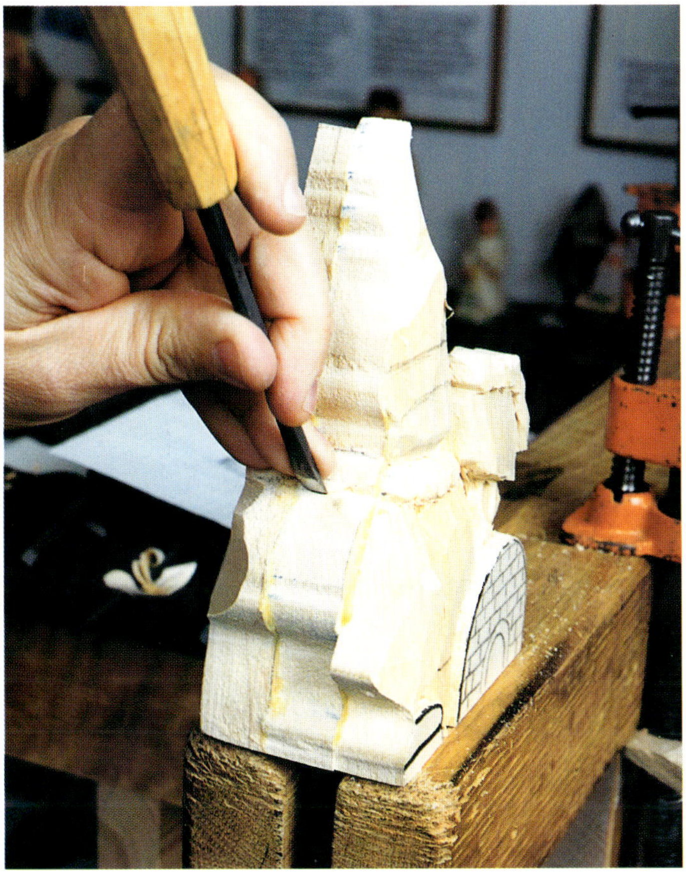

Use a small "U" gouge to outline the bottom of the scarf.

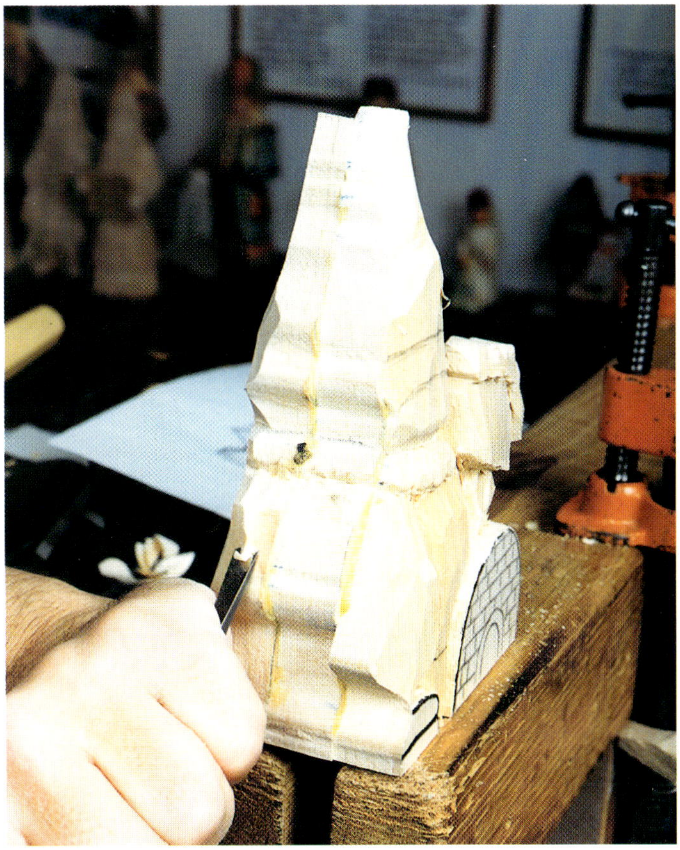

Round the shoulder.

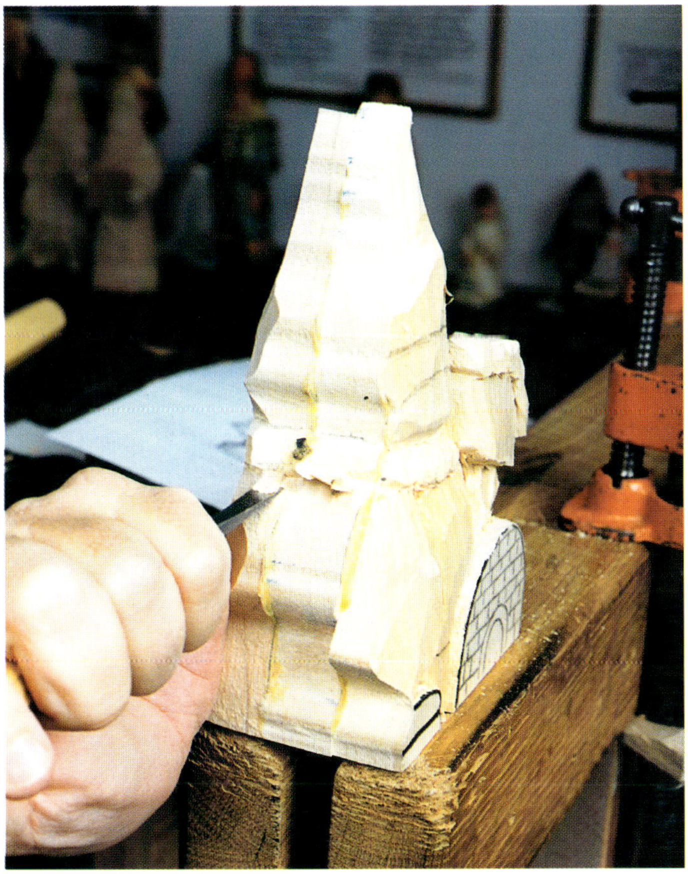

Use the small "U" gouge to cut up into the stop cut and reveal the scarf at the side of the neck.

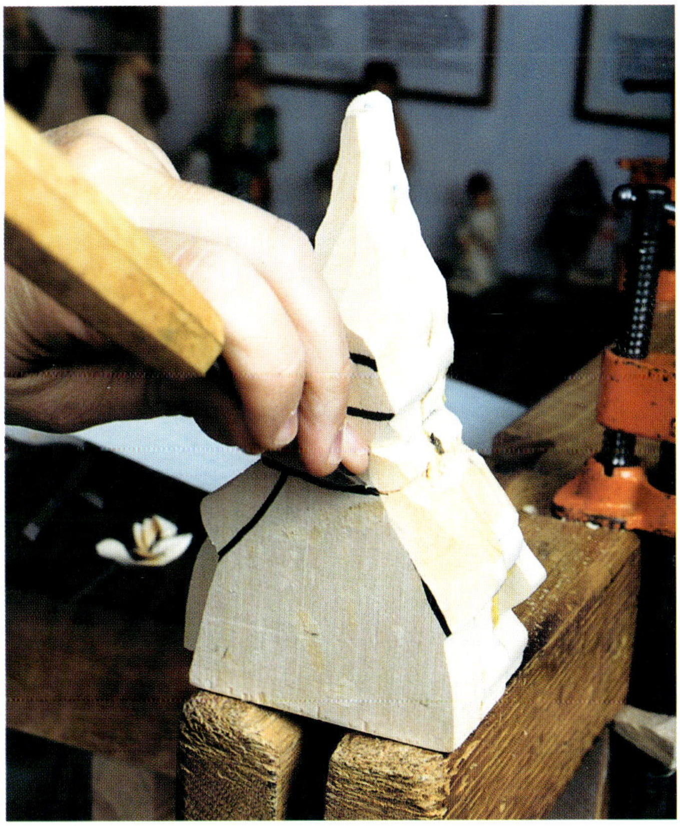

Now I am working on the back of the piece. As you see, the process of carving is like peeling an onion – you go layer by layer. Don't go too far in one area since you may throw the piece out of balance. Going slowly, layer by layer, will help keep the piece in balance.

Start the back by outlining the bottom of the scarf with a stop cut.

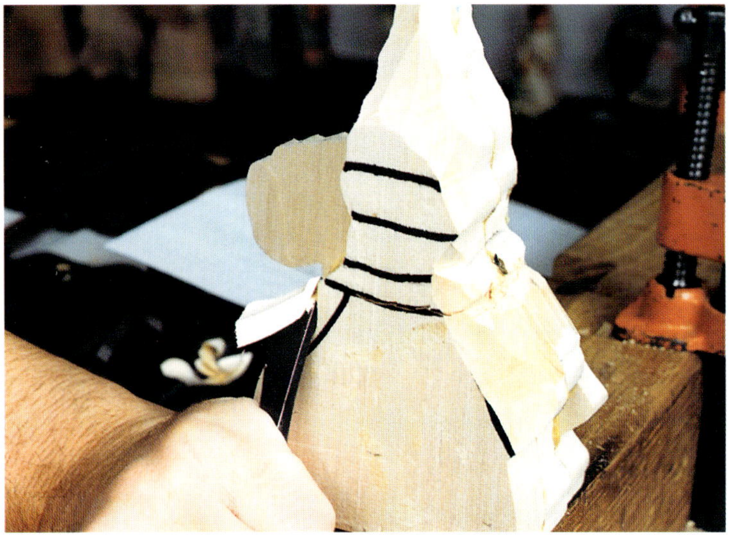

Use a large shallow "U" gouge to round the shoulder.

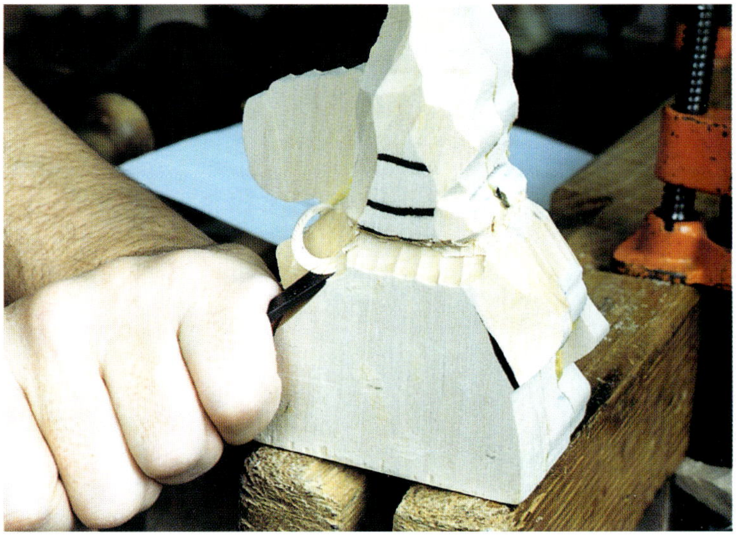

Remove more wood and keep rounding the back.

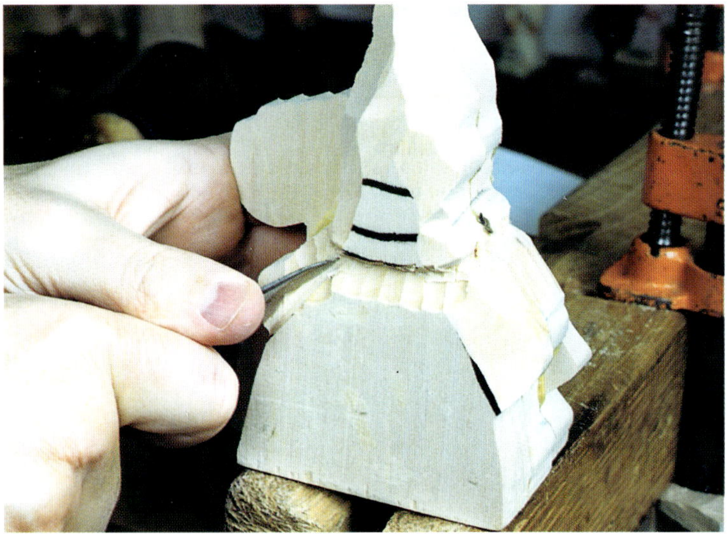

Use a knife to cut the line at the back of the Snowman's arm. Note that I used a large shallow "U" gouge to begin rounding the back of the hat too.

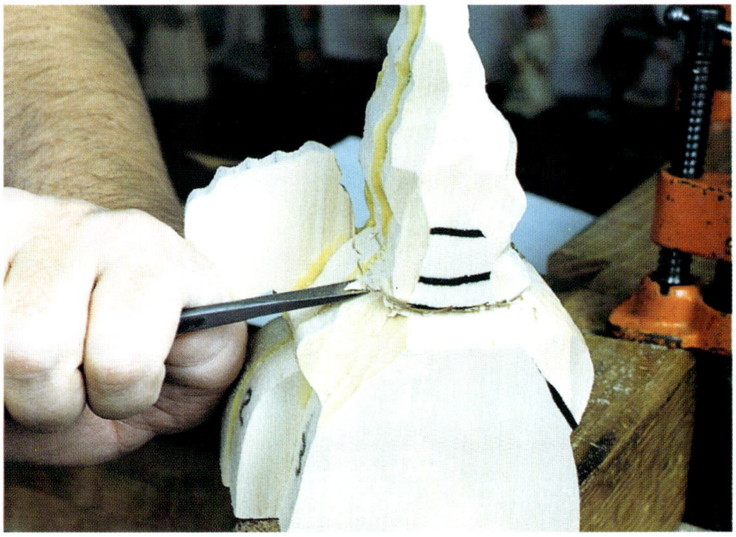

Here I have moved to the last side of the Snowman. I am moving around him "peeling the onion" as I go.

Make a stop cut at the bottom of the scarf and use a small "U" gouge to cut up and under the scarf.

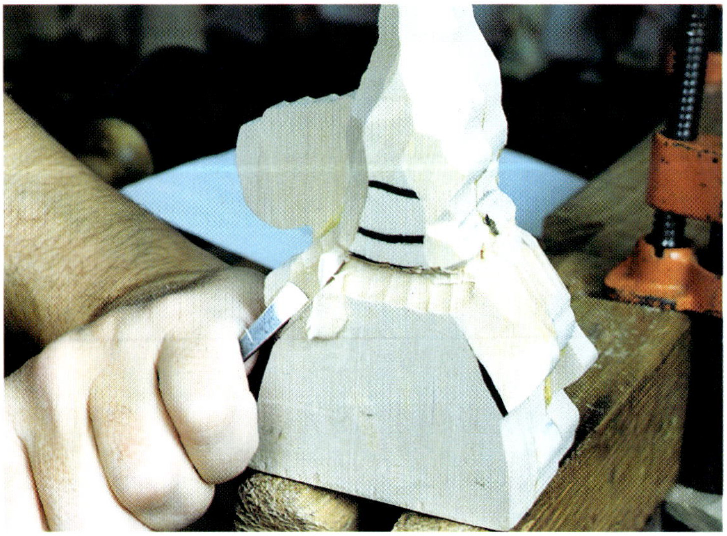

Use a small shallow "U" gouge to remove wood from the back into the stop cut.

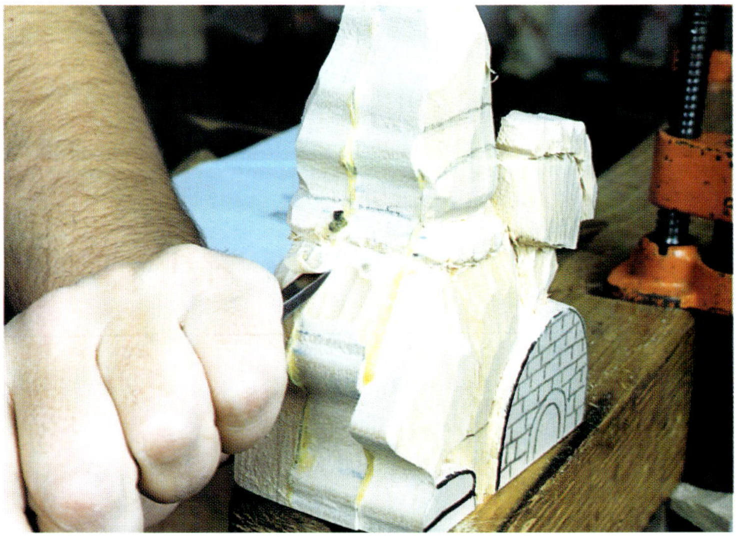

Move around the piece rounding up into the scarf. Step back from the piece and examine it so that you keep it in perspective.

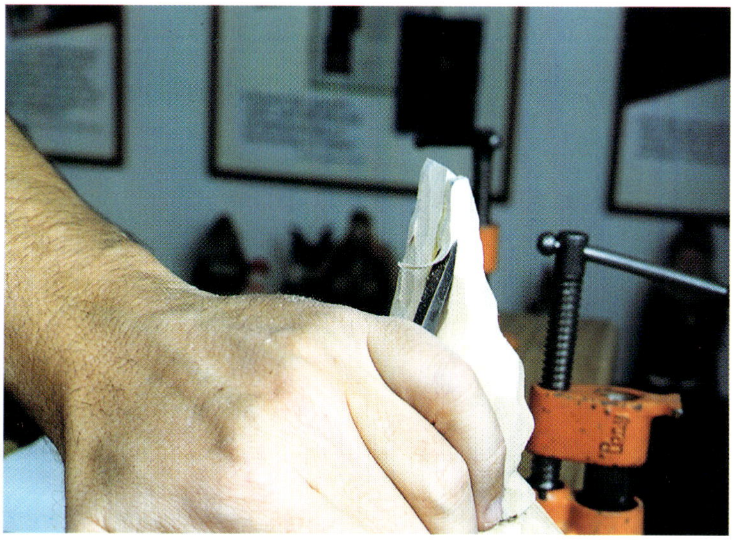

Use a larger shallow "U" gouge to shape the hat. The shape is up to you – just don't cut too much, too fast.

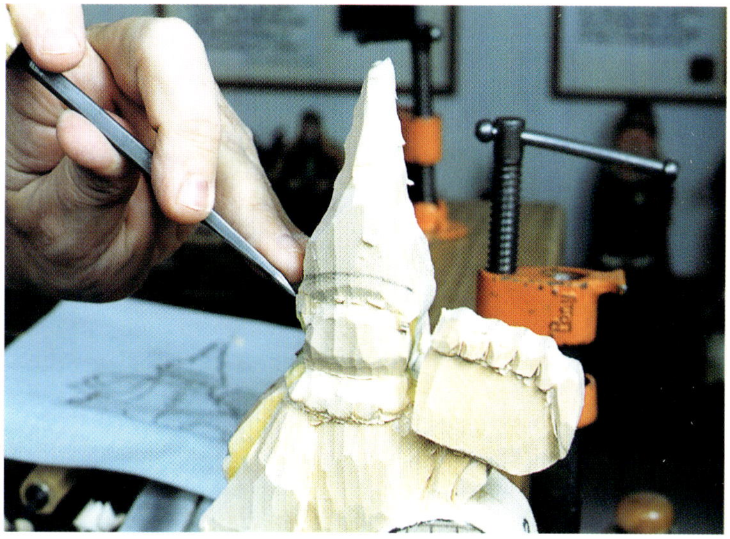

Move around the hat bottom establishing the line between the hat and head.

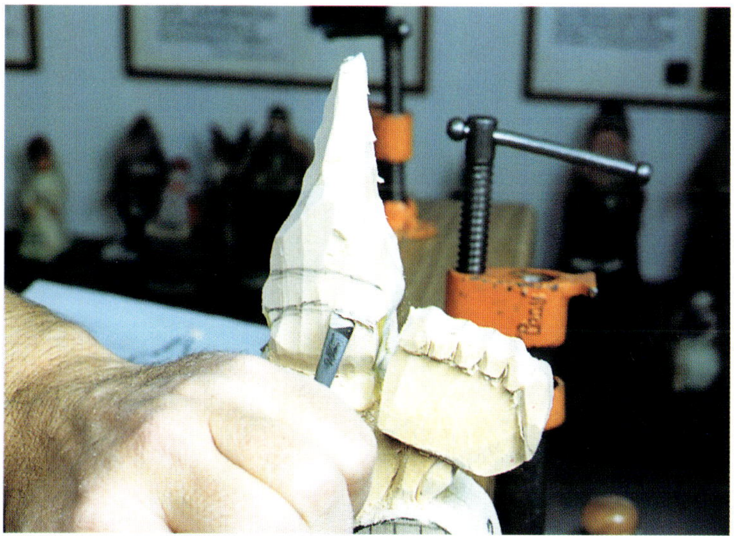

Make a stop cut along the bottom of the stocking cap. Cut the face back using a small "U" gouge to shape the face.

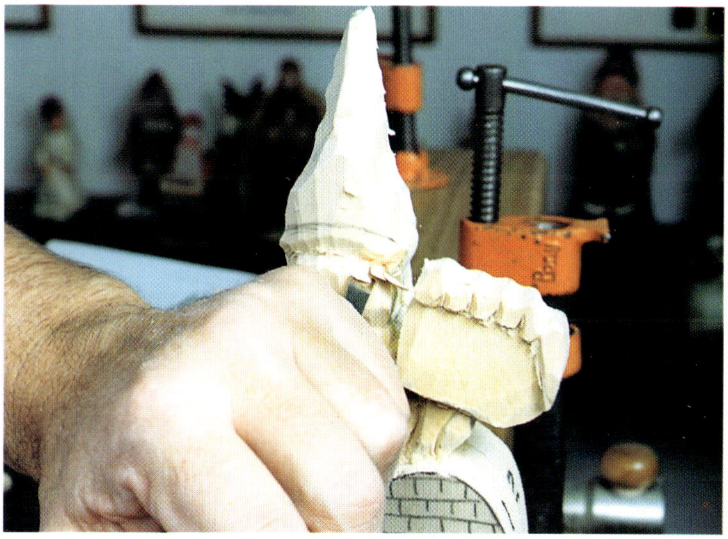

Shape the head on the right side.

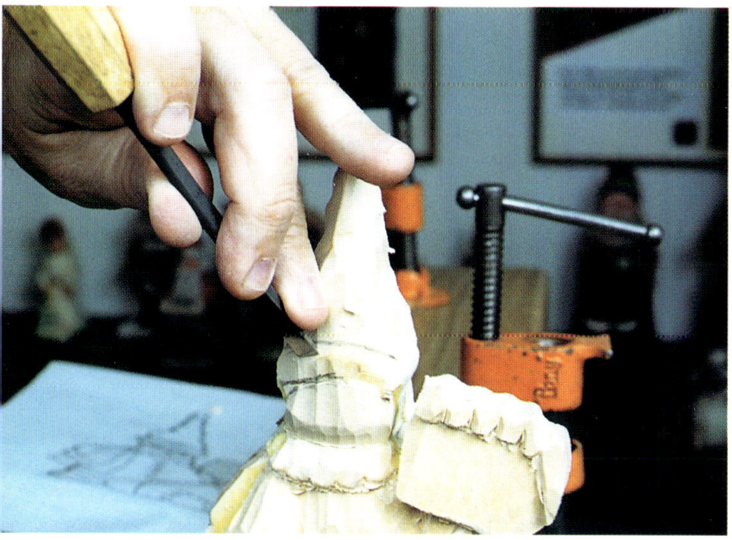

Use the gouge to cut down the hat bottom into the stop cut. This will shape the bottom of the stocking of the hat.

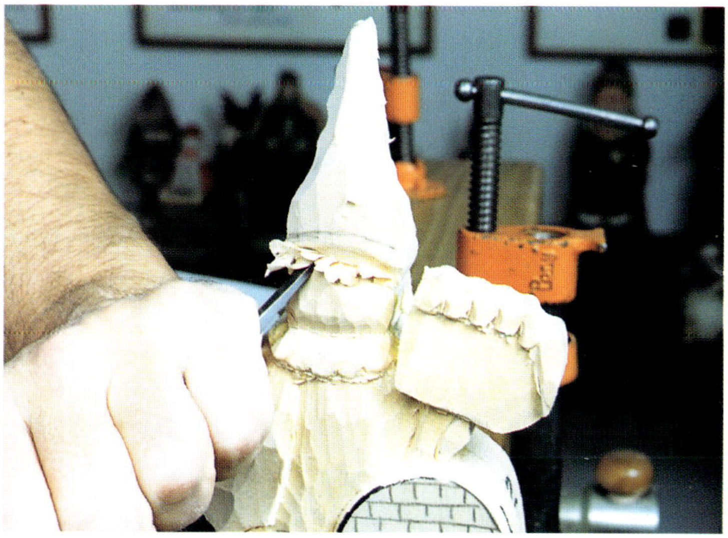

Deepen the area at the bottom of the hat.

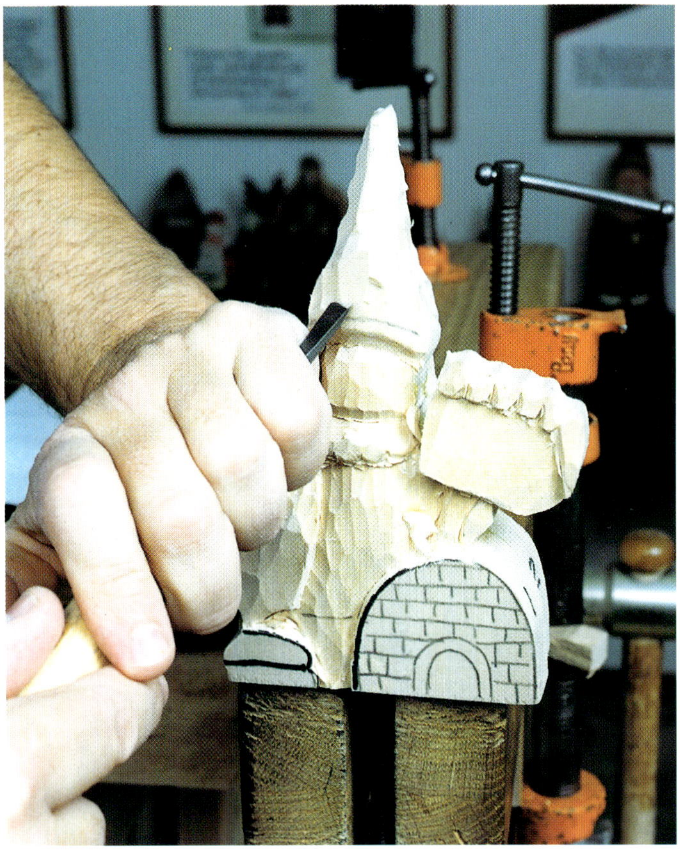

Shape the hat back – the actual shape is up to you, so think about it a little before you cut. I like a hat that looks "wavy" or "slouchy" as if it has settled down on his head.

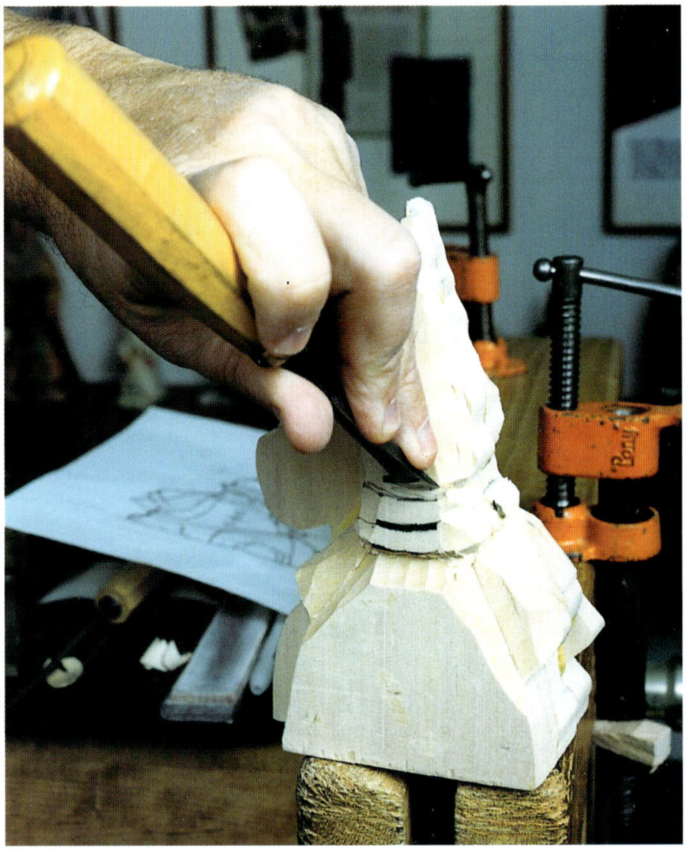

Work on the back of the piece cutting in the bottom of the hat and the top of the scarf.

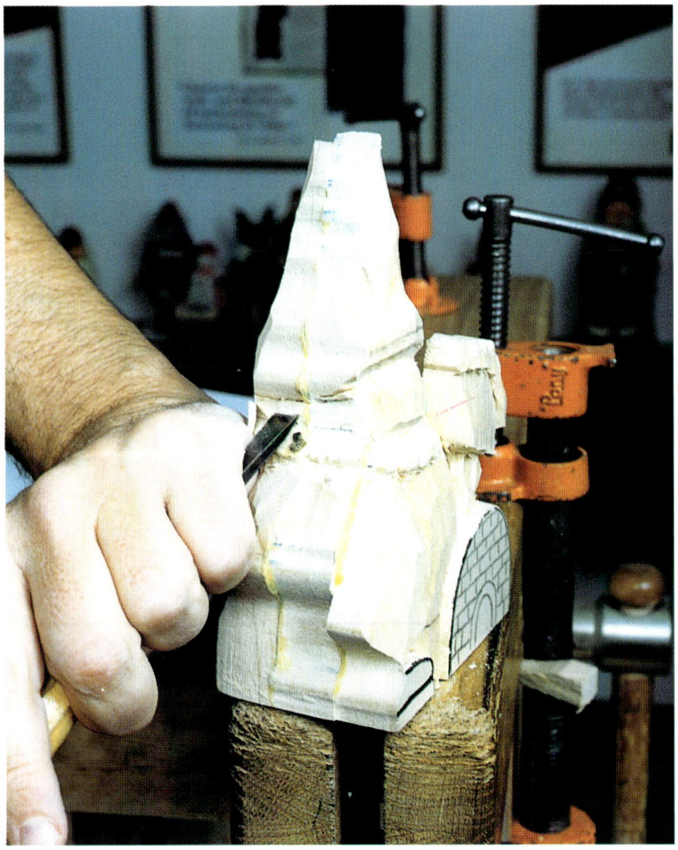

Establish the hat – head separation all the way around the carving.

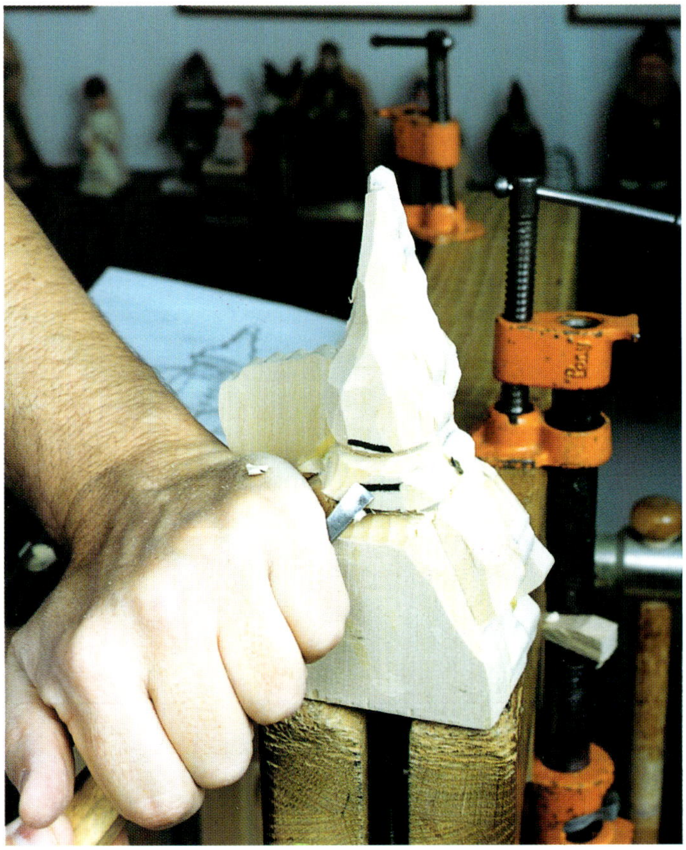

Work on the scarf.

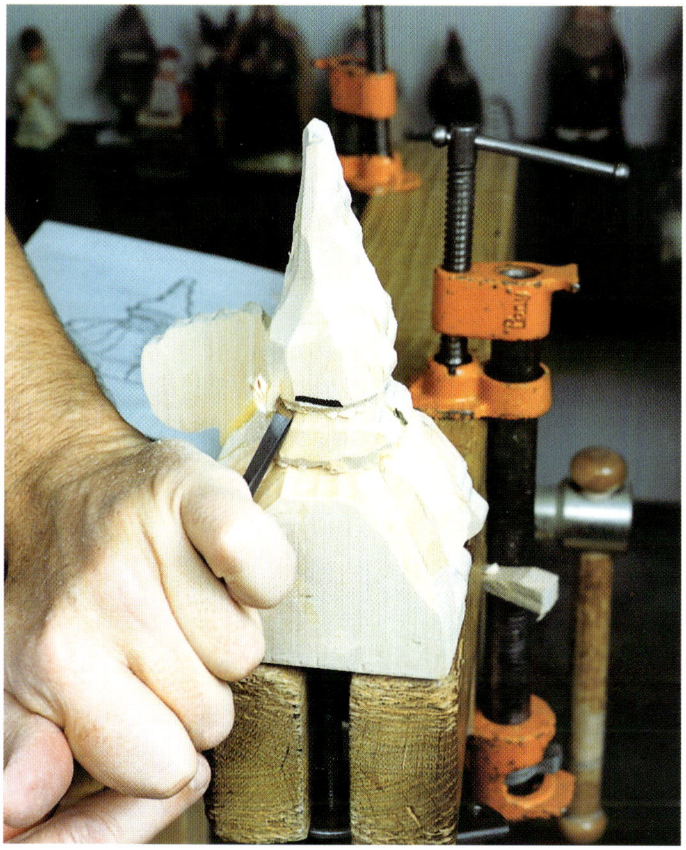

Shape the back of the hat.

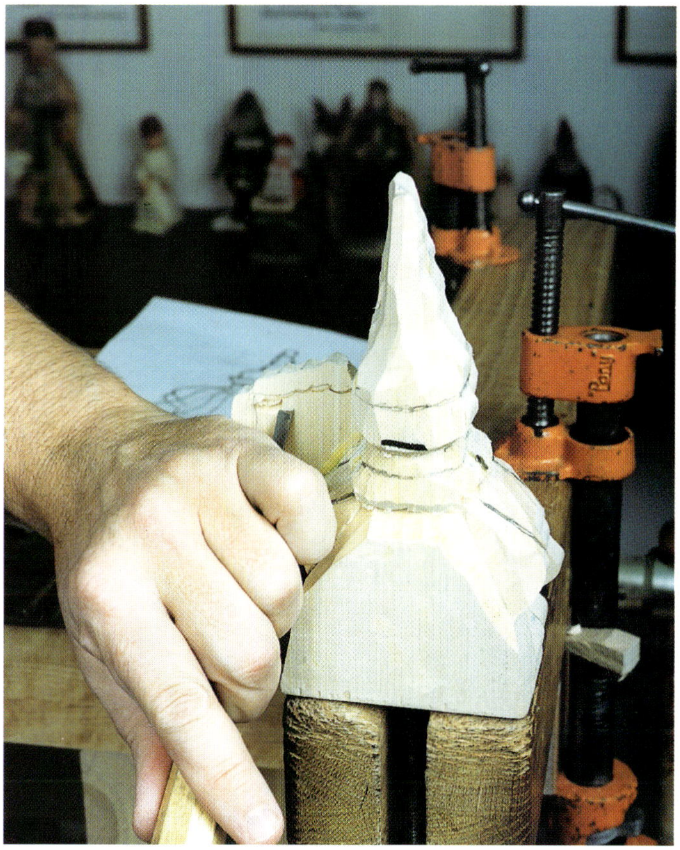

Cut up into the stop cut to help the snow stand out more.

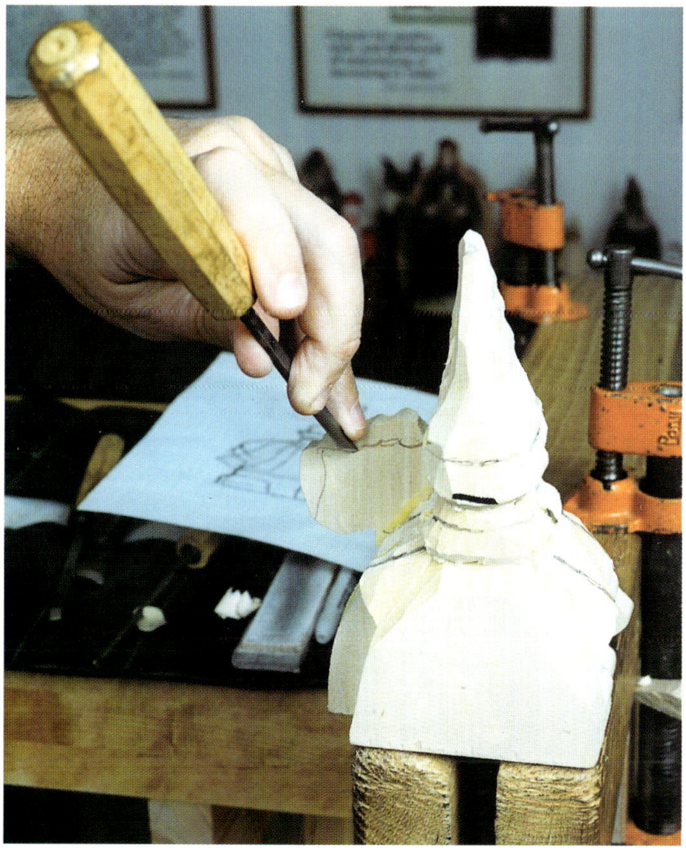

Make a stop cut on the back of the sign to define the "snow" at the top and on the side of the sign.

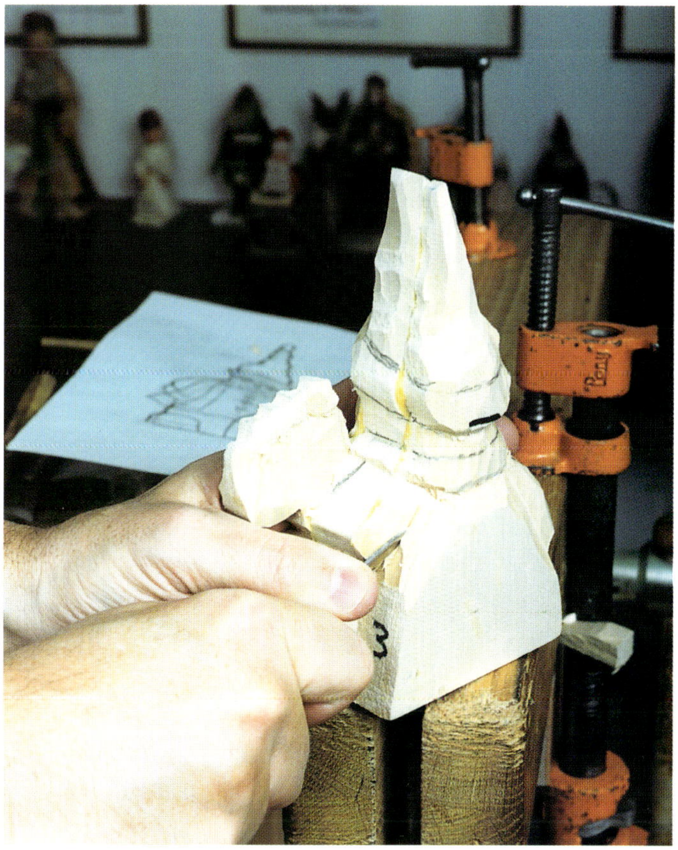

Deepen the cut defining the back of the Snowman's arm.

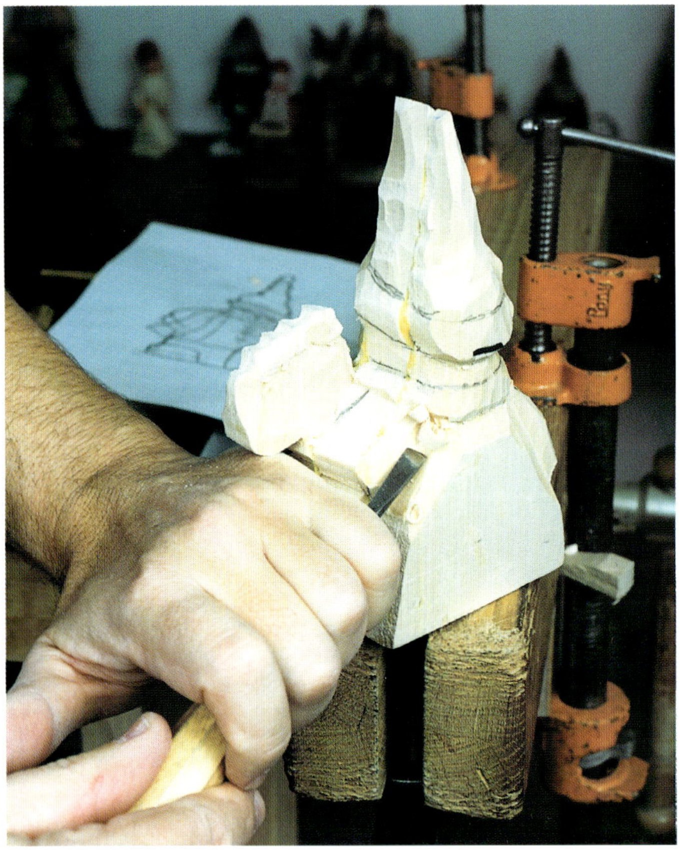

Use a small "U" gouge to remove wood from behind the arm.

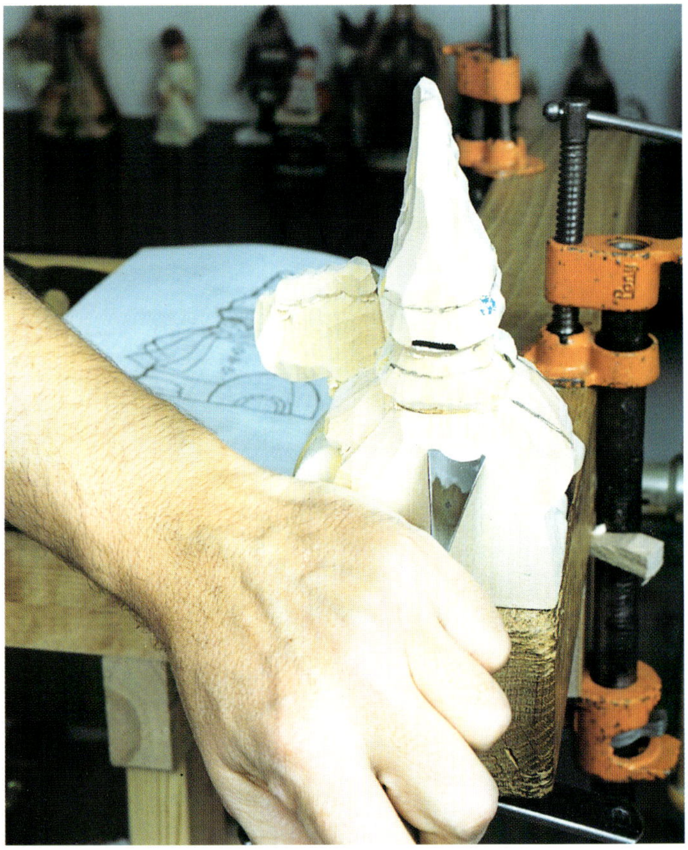

Use the large shallow "U" gouge to shape the rest of the back.

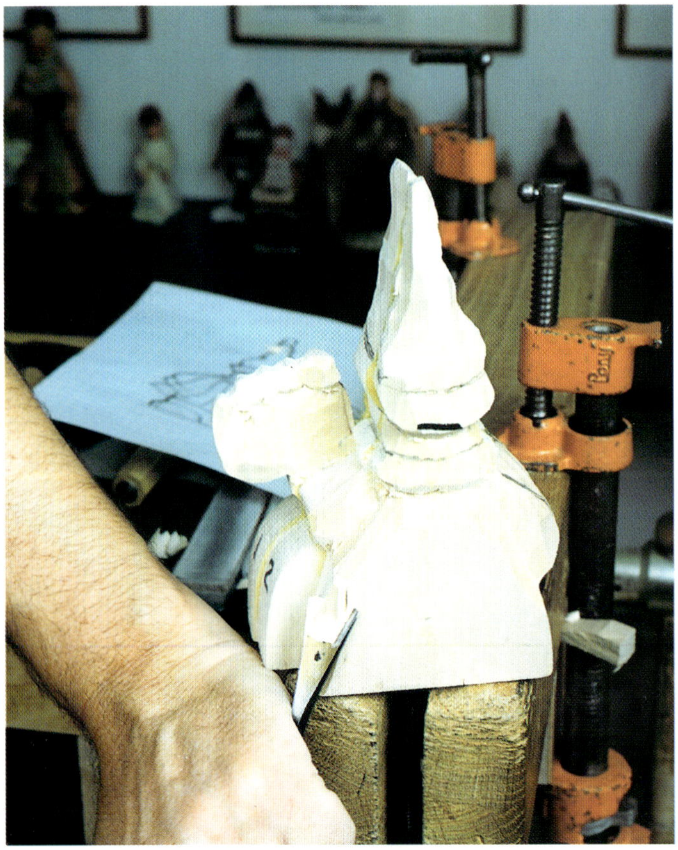

Use a large shallow "U" gouge to shape the bottom back of the Snowman.

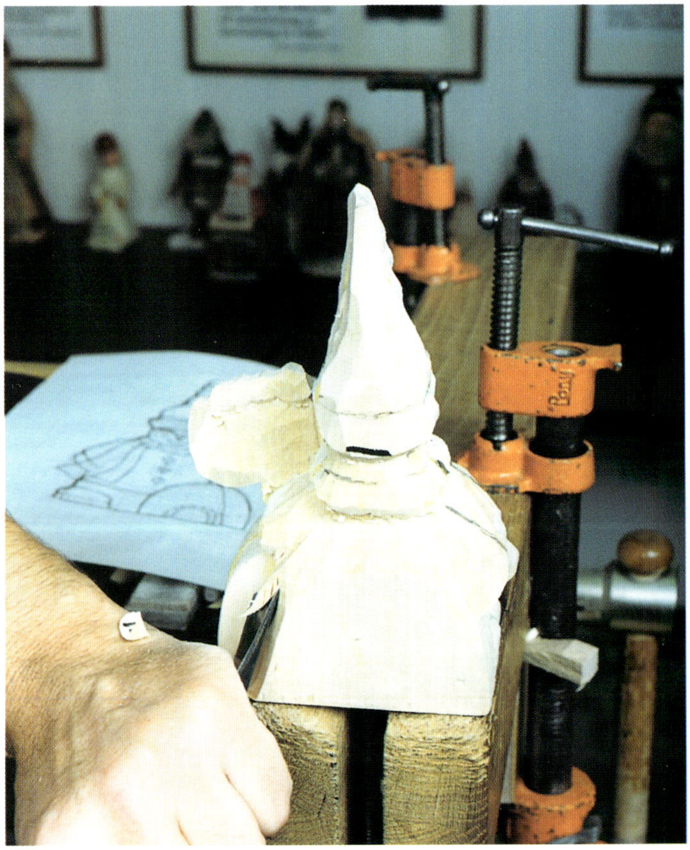

Continue the shaping.

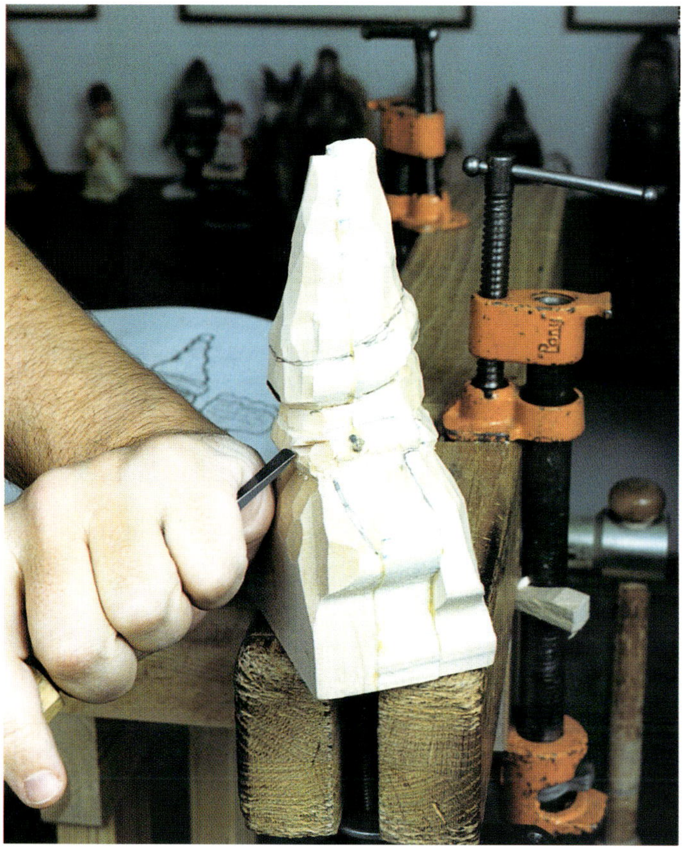

This is a good picture to examine. It shows that while we have made progress, the piece is still too fat and there are areas that have not been well defined --like the right arm.

Shape the areas that are too fat – use your own judgment on how he looks.

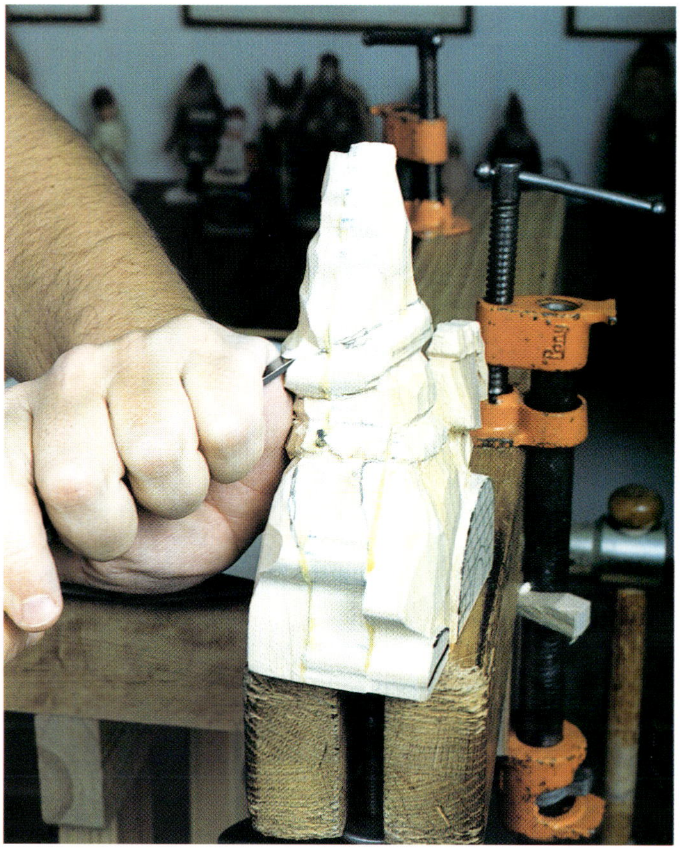

Shape the hat.

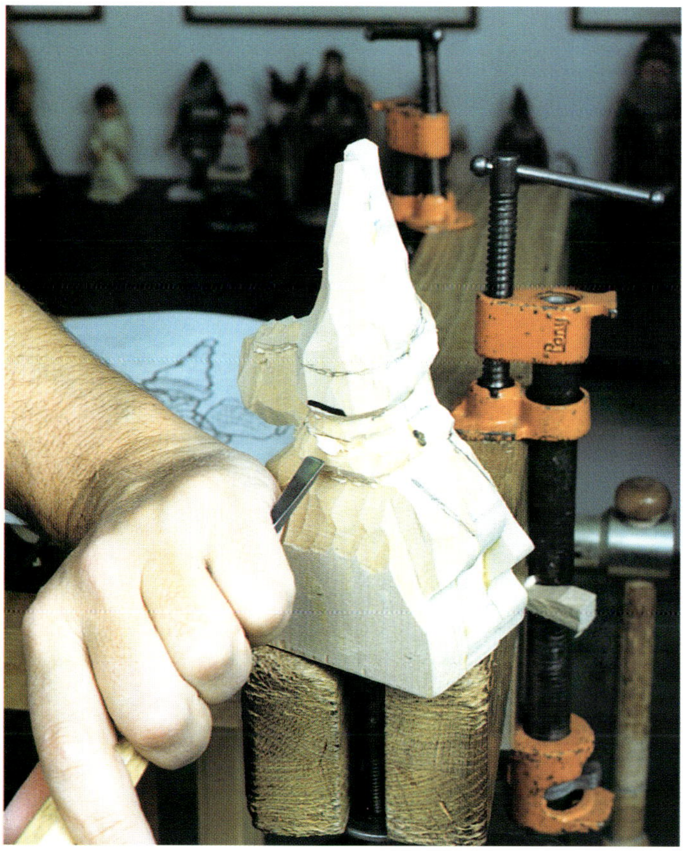

Continue to shape.

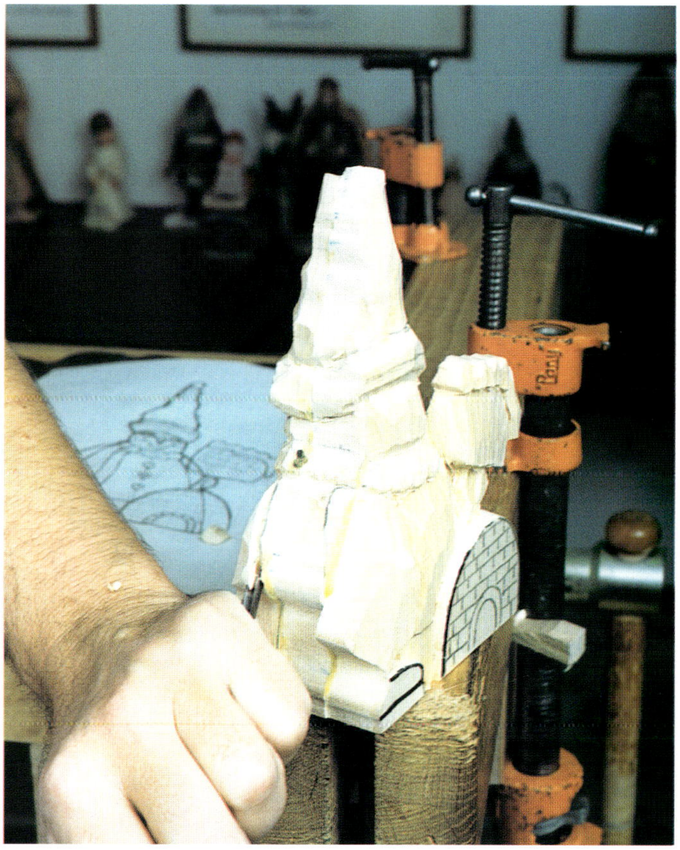

Outline the back of the right arm and then remove some of the wood behind the arm.

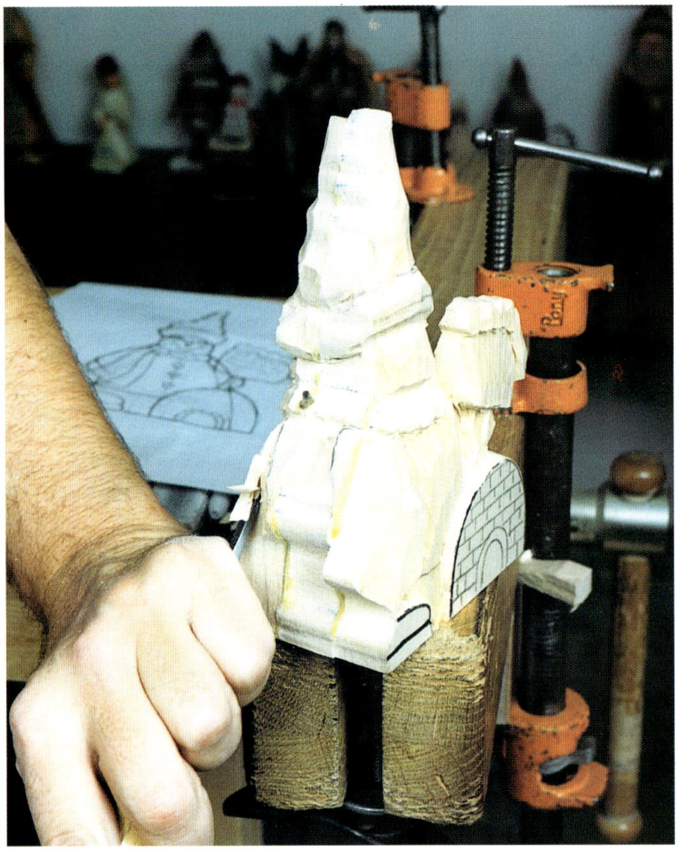

Remove wood from behind the arm and shape the back.

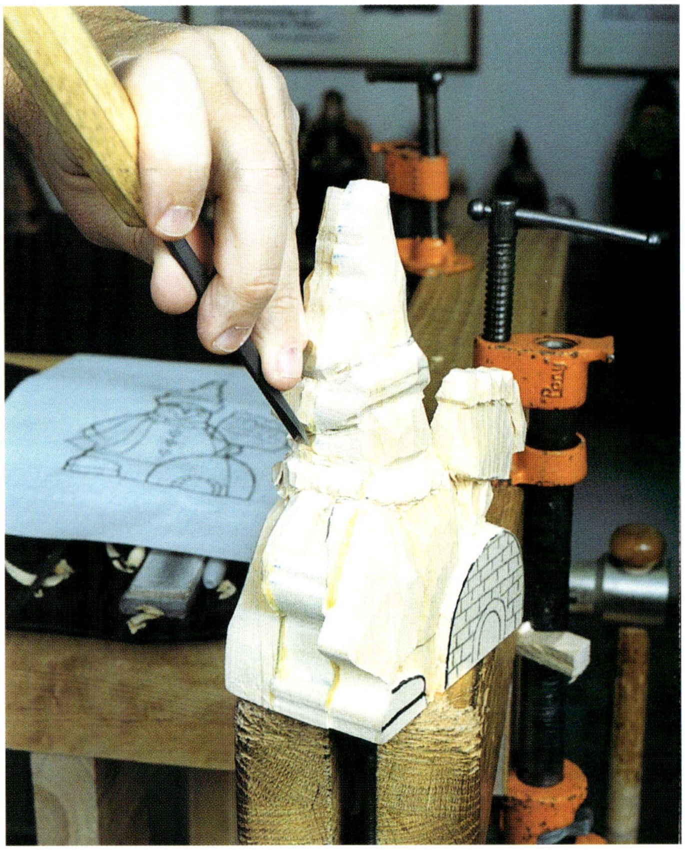

Make additional stop cuts as needed to help shrink the size of the head and define the carving.

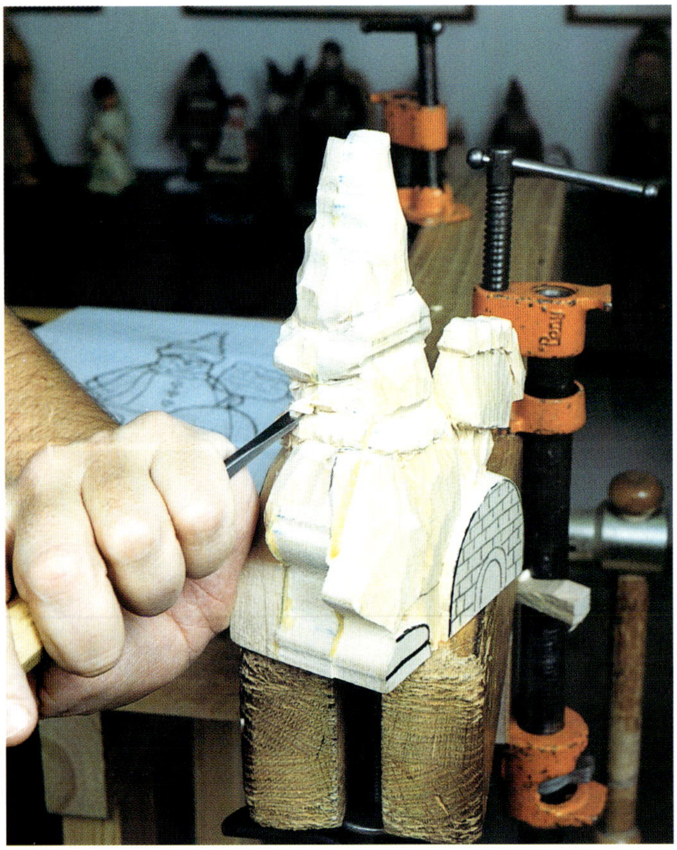

Shape the head and the scarf.

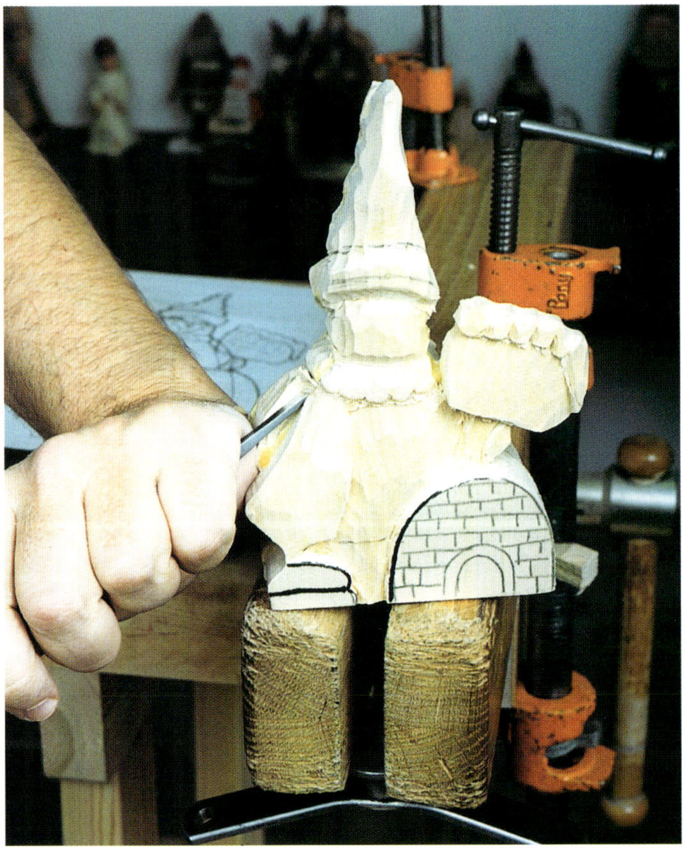

The Snowman is starting to look pretty good here but he still needs some more adjustment and defining. Here I am working on the left side of the scarf to shape the arm and make the scarf look more like it is blowing in the wind.

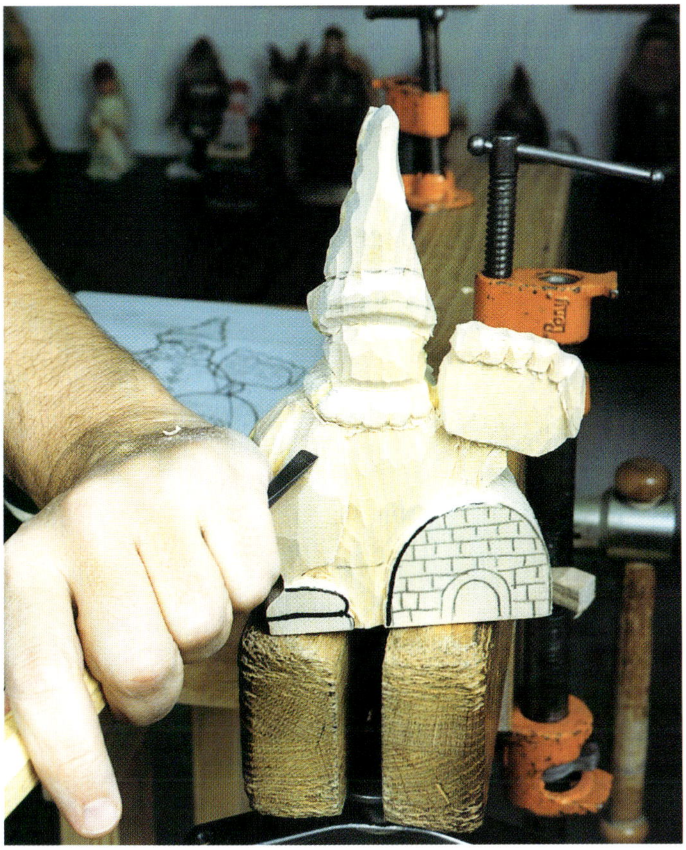

Shape the scarf.

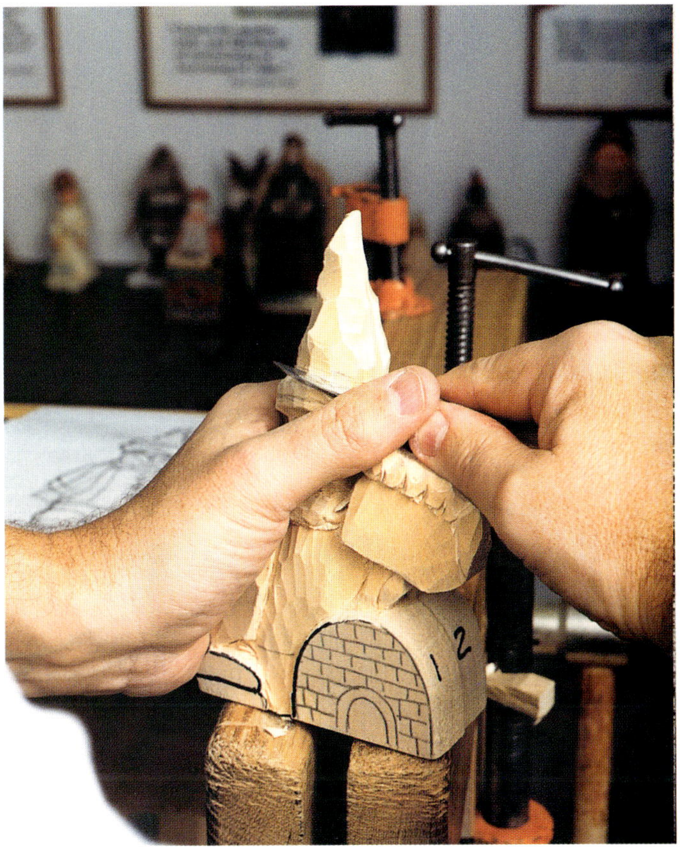

Here I am using a knife to shape the hat. You can use a gouge just as well.

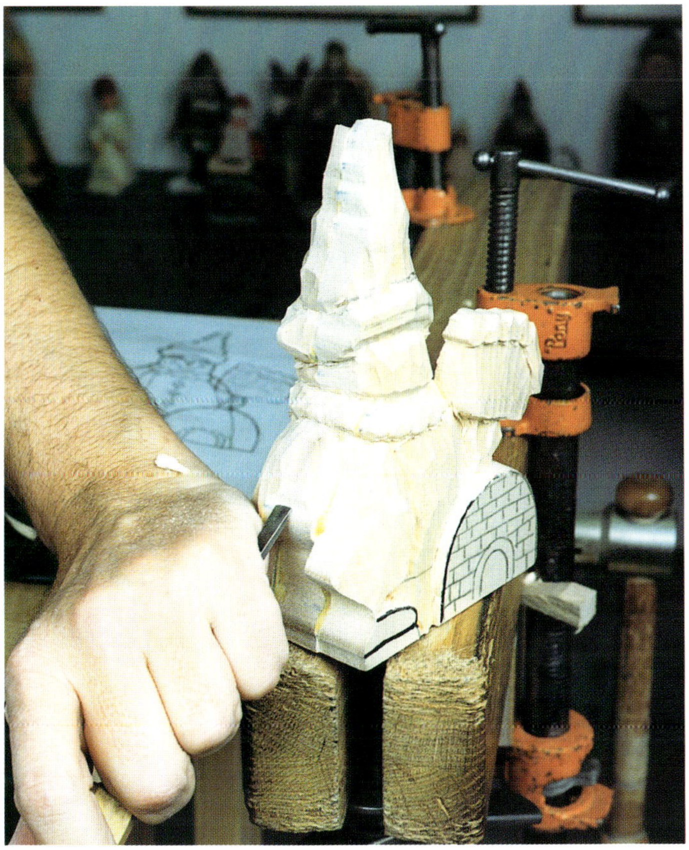

Cut a line separating the Snowman's hand from his sleeve.

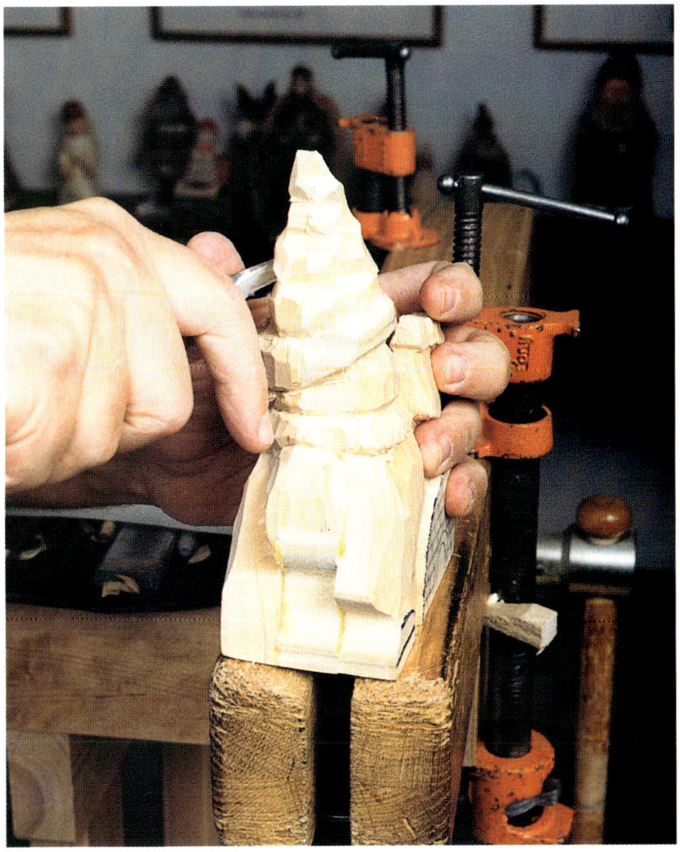

Shape the hat.

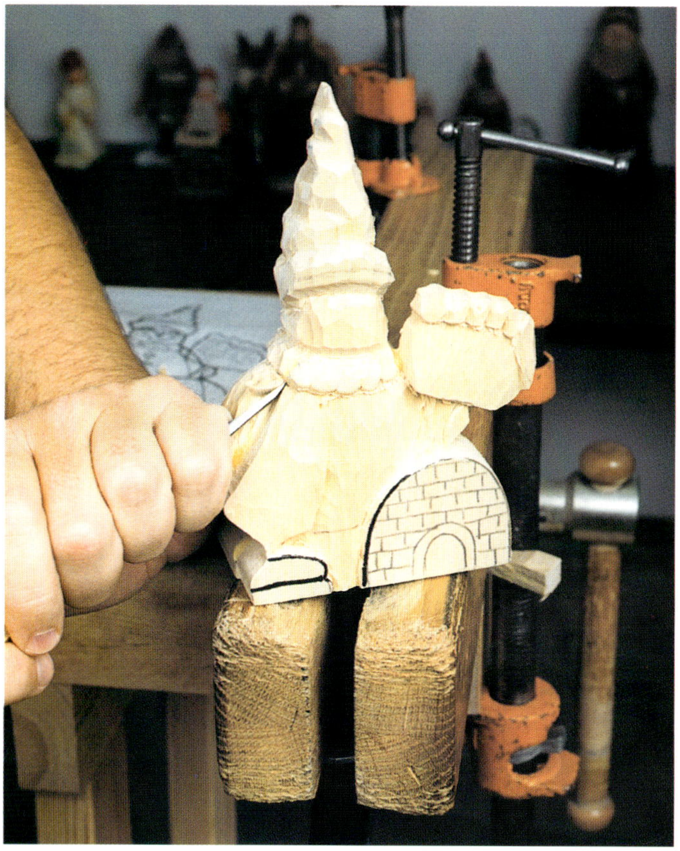

Define the arm and scarf. Cut away the scarf to reveal more arm.

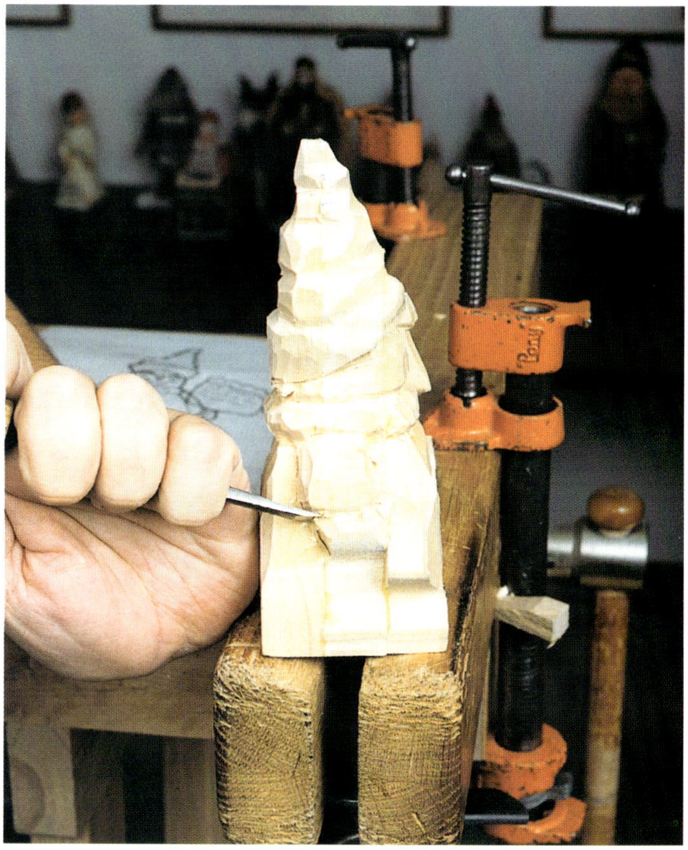

Cut the line defining the bottom of the sleeve.

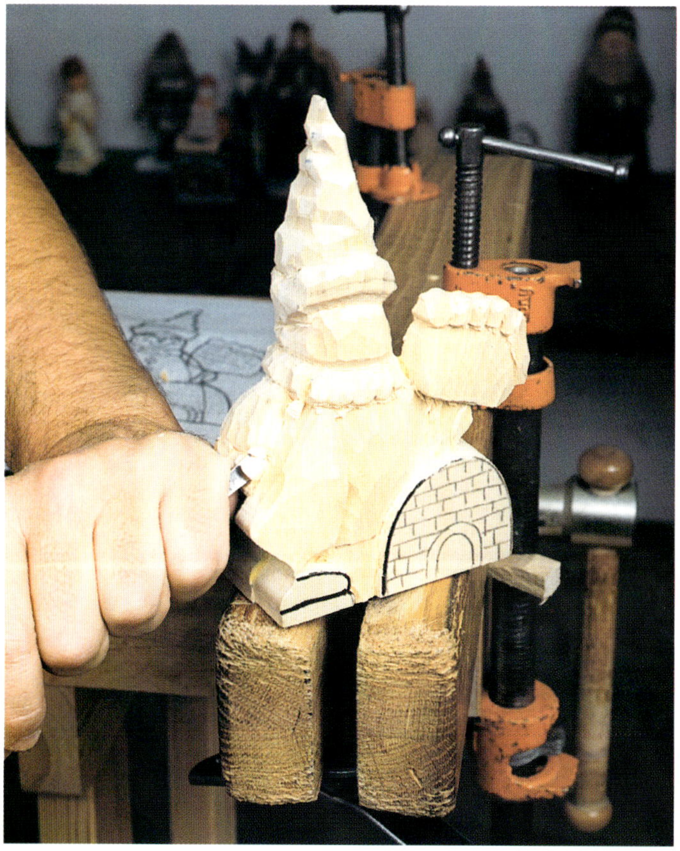

Shape the body below the arm.

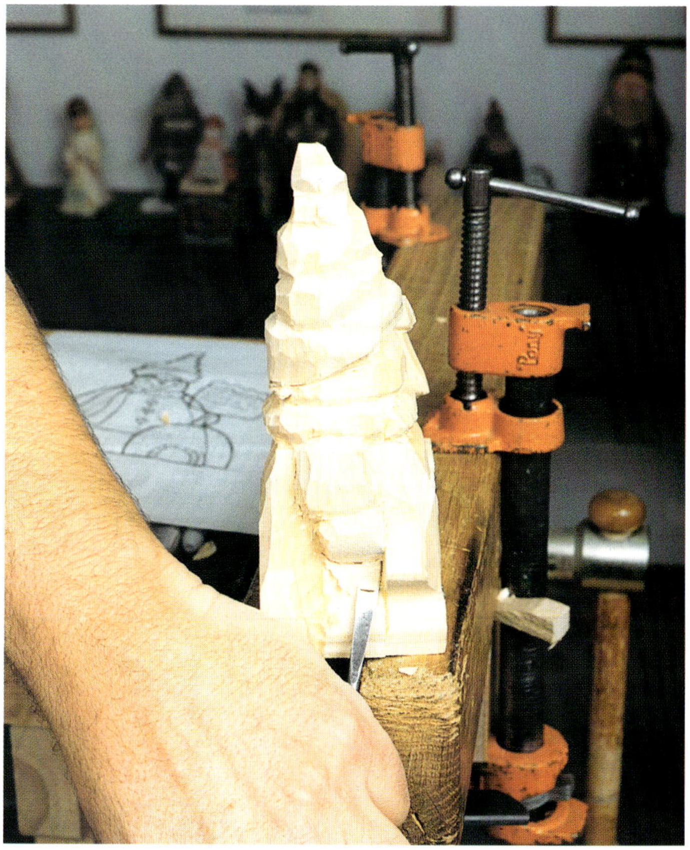

Shape the body below the arm.

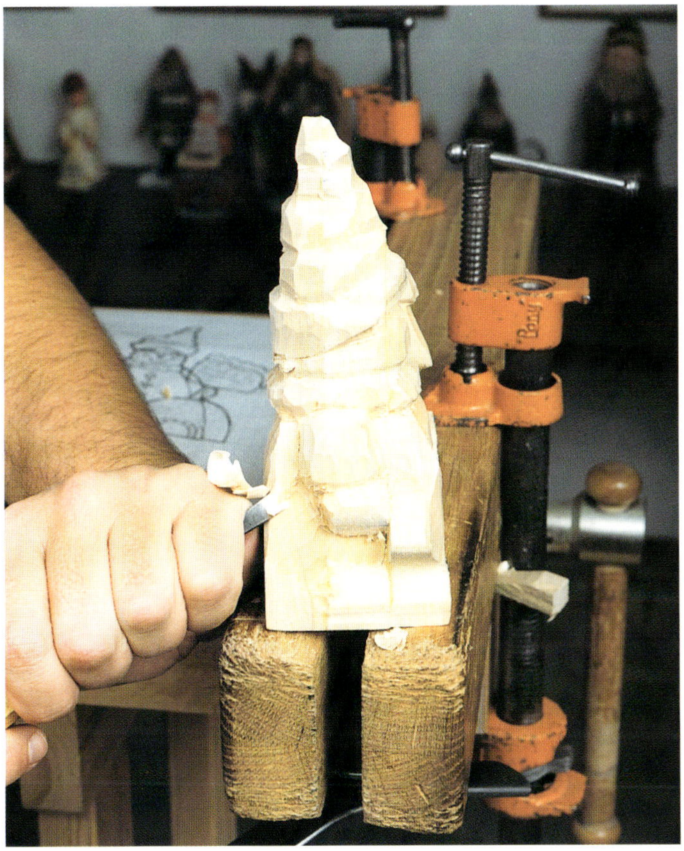

Shape the back of the body. Look at this picture closely and see that the body still needs to be shaped.

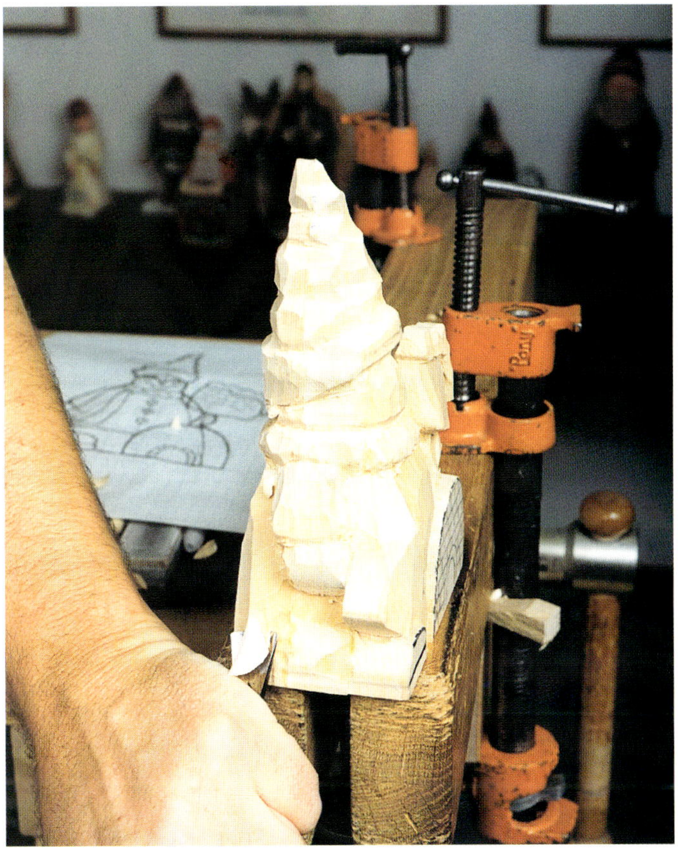

Use a large shallow "U" gouge to shape the bottom of the body.

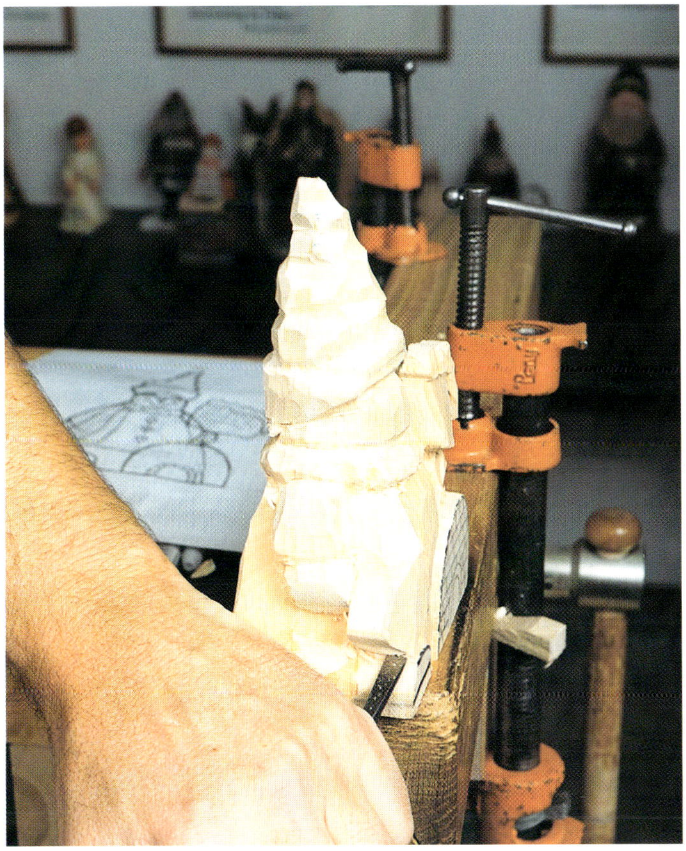

Define the top of the boot and separate it from the bottom of the scarf.

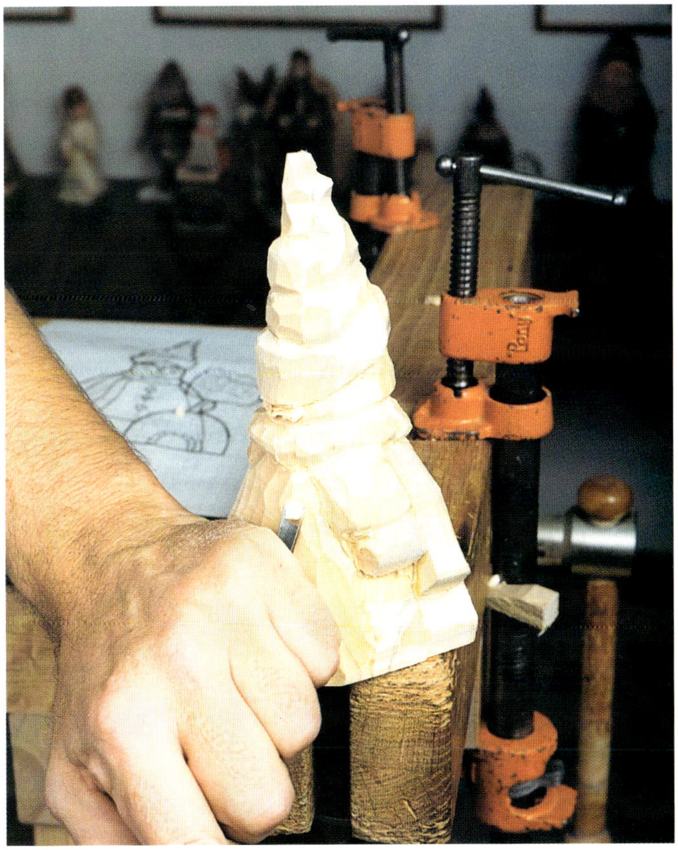

Shape the back of the body.

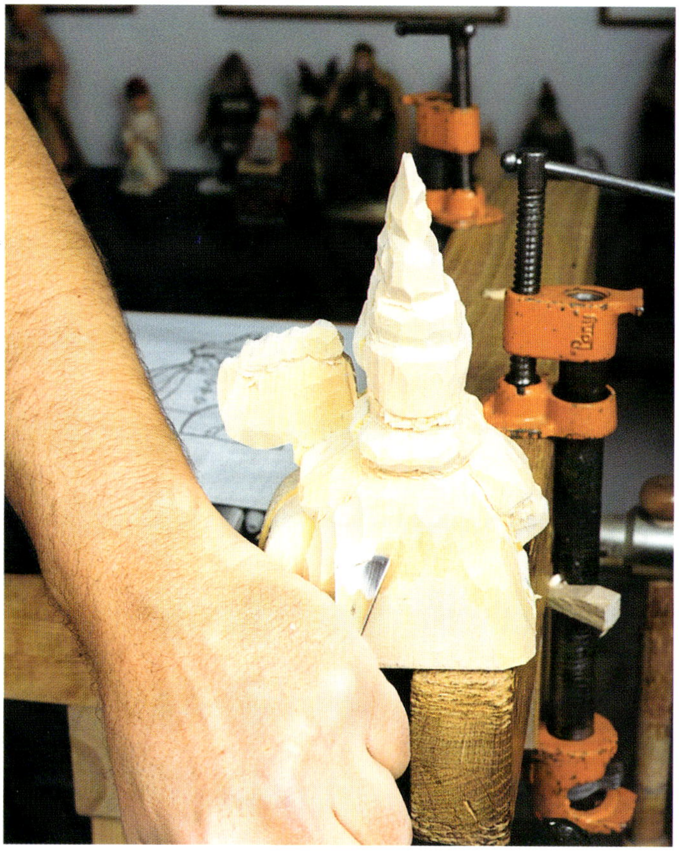

Use the large shallow "U" gouge to shape the back of the body.

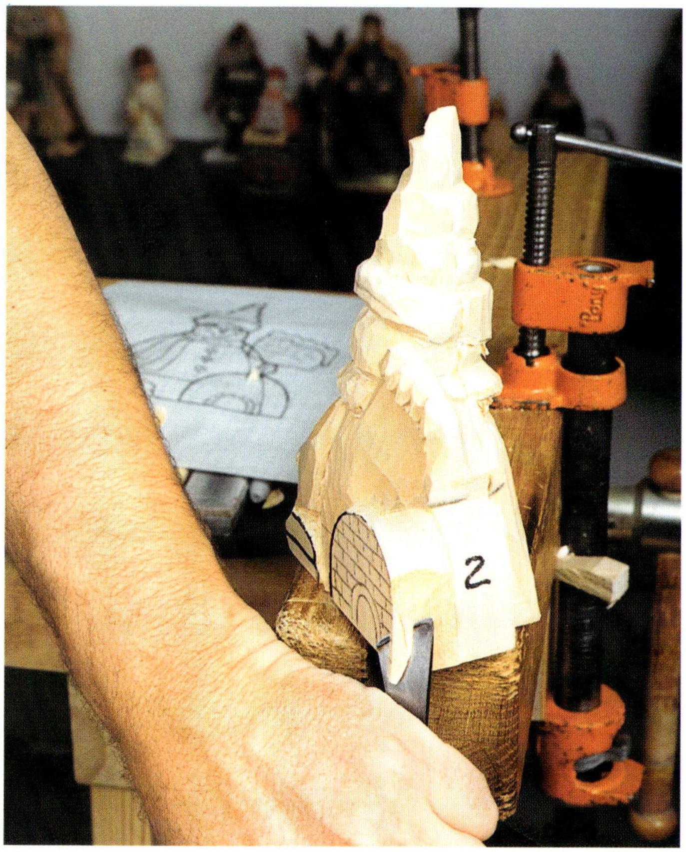

Use the large shallow "U" gouge to start shaping the front of the igloo.

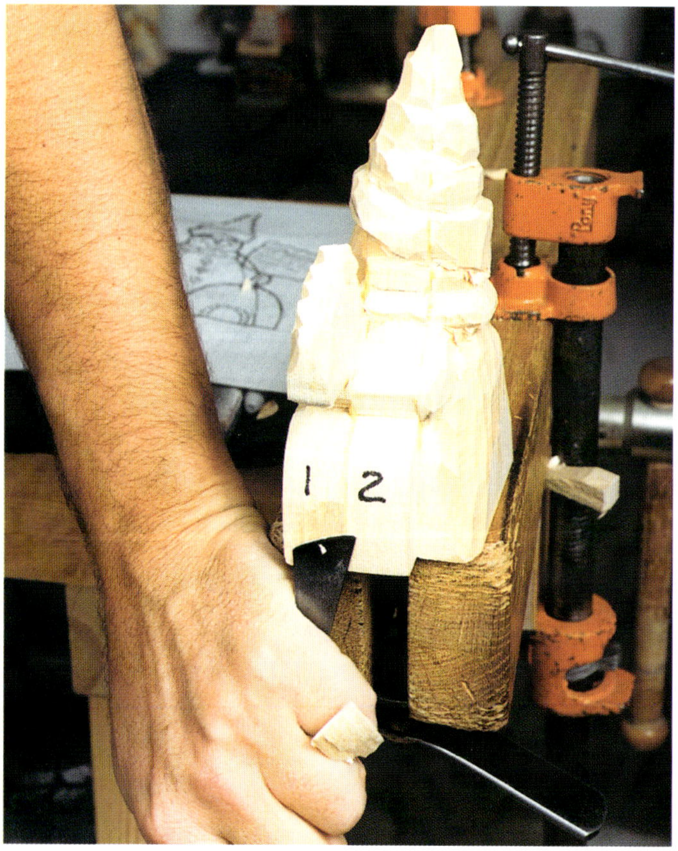

Use a large shallow "U" gouge to begin shaping the igloo. You can see that we have not touched this area at all yet.

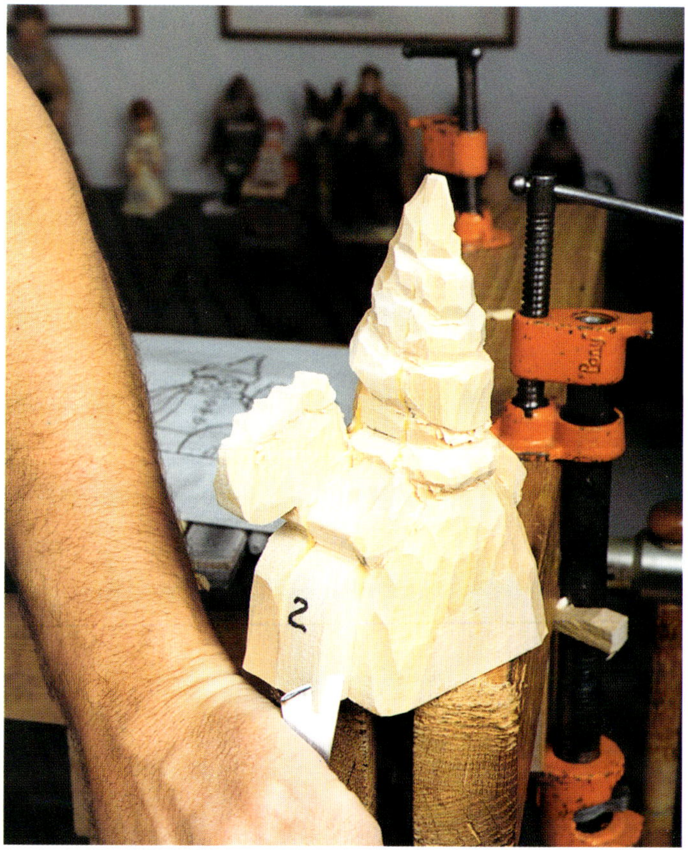

Continue to shape the igloo.

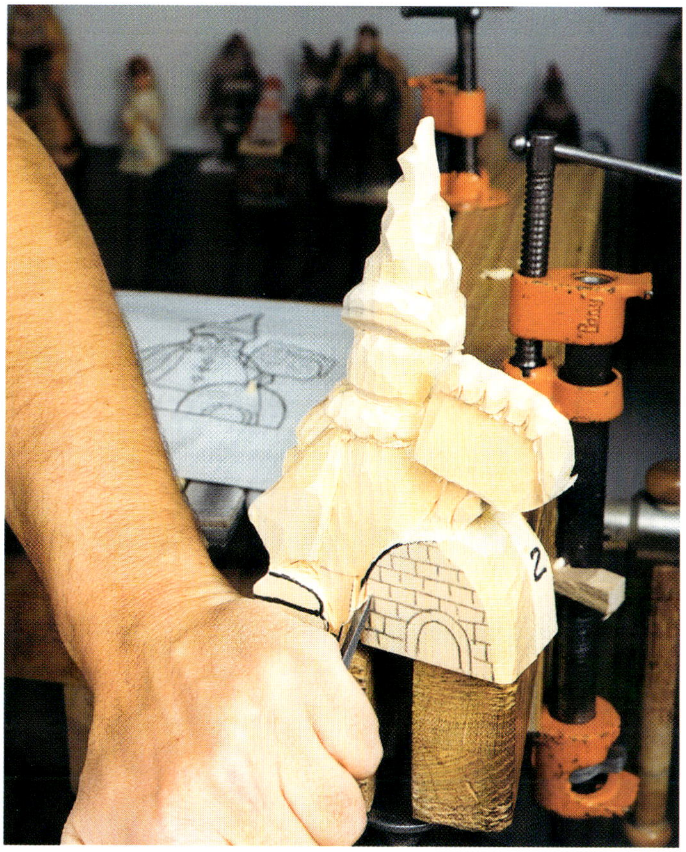

Shape the left side of the igloo.

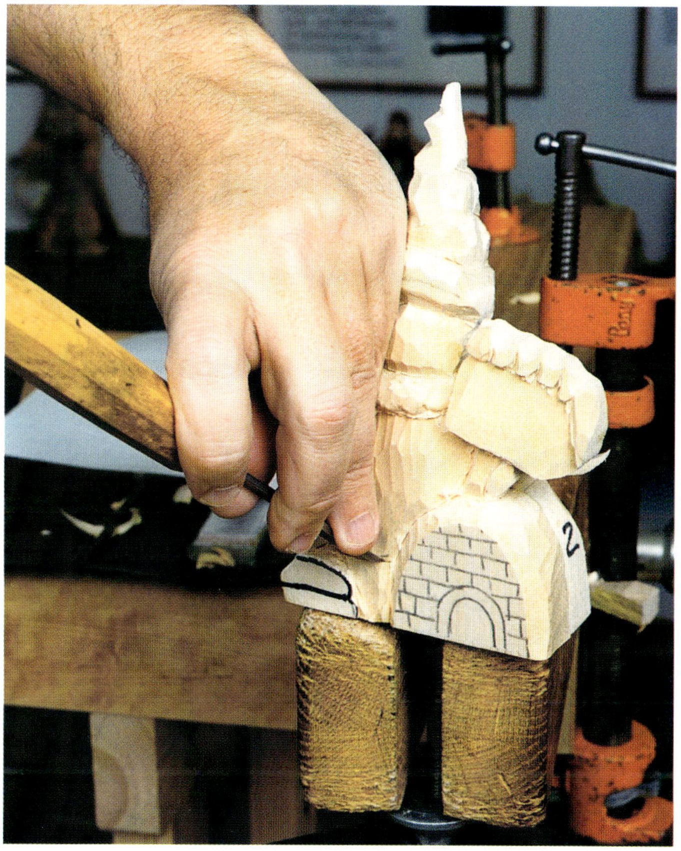

Cut a stop cut at the bottom of the Snowman's sweater.

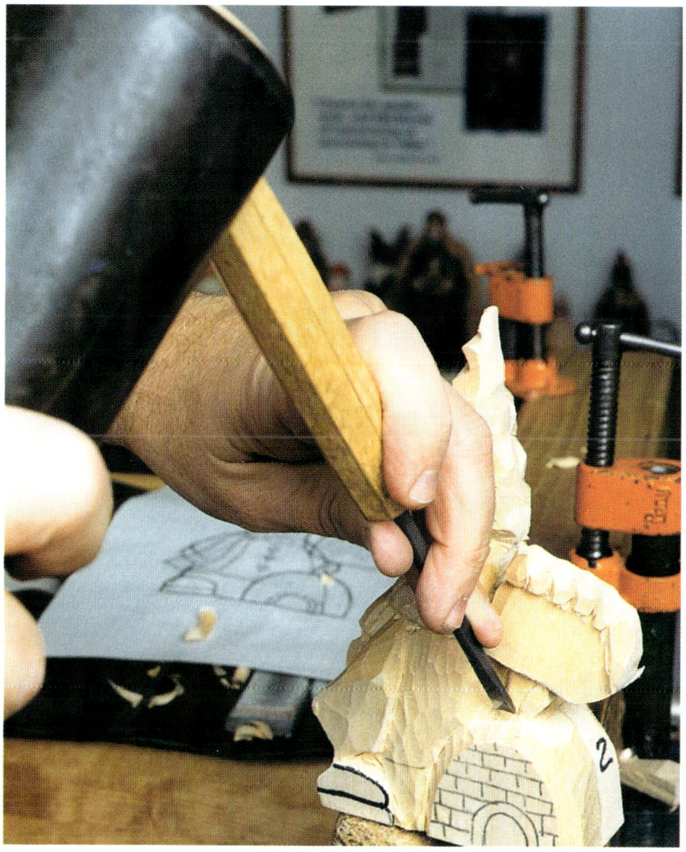

Cut a stop cut at the bottom of the "stick" holding up the sign.

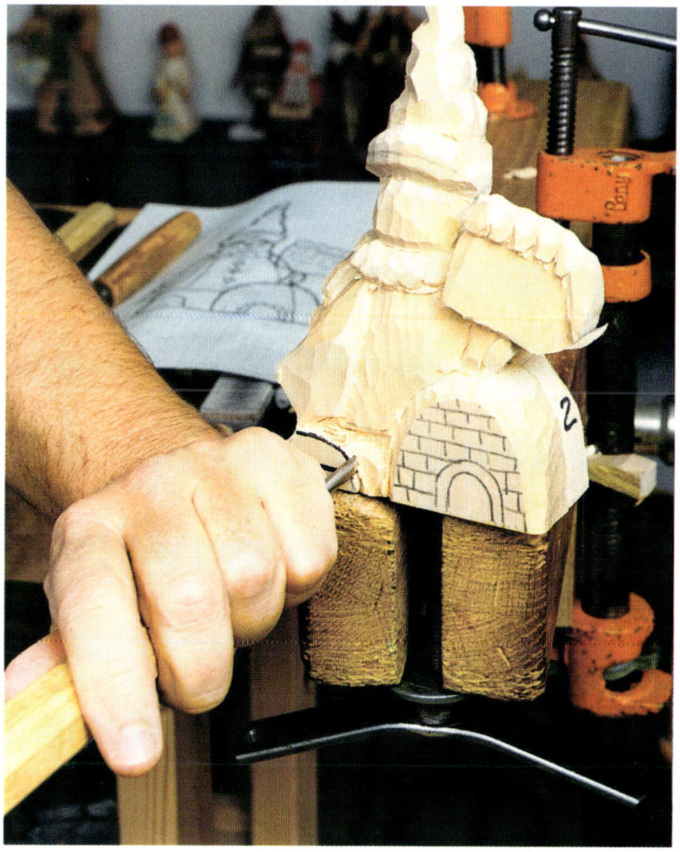

Cut up into the stop cut and down into the stop cut to define the bottom of the sweater. Re-cut the line at the edge of the boot.

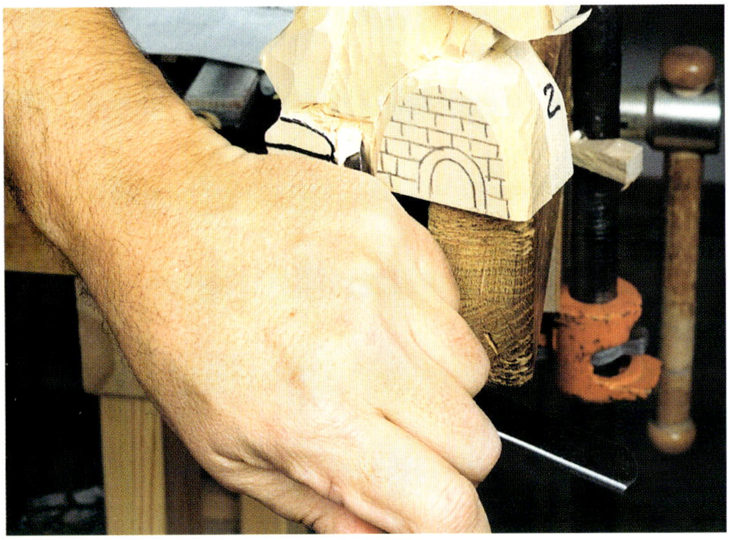

Remove wood from under the sweater and inside of the boot to help define the boot.

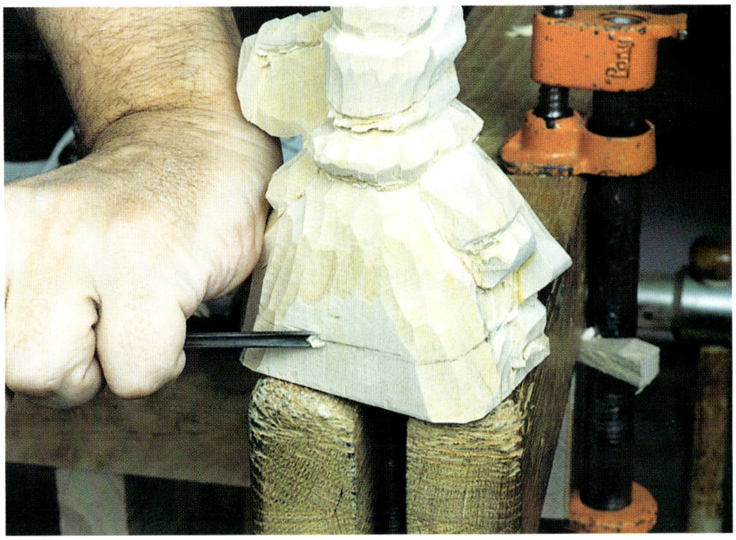

Use a "V" gouge to cut the line at the bottom of the sweater.

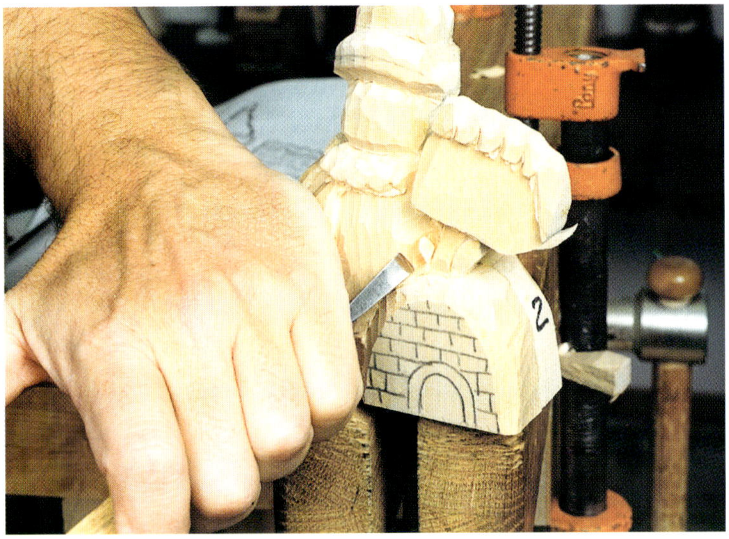

Shape the top of the igloo.

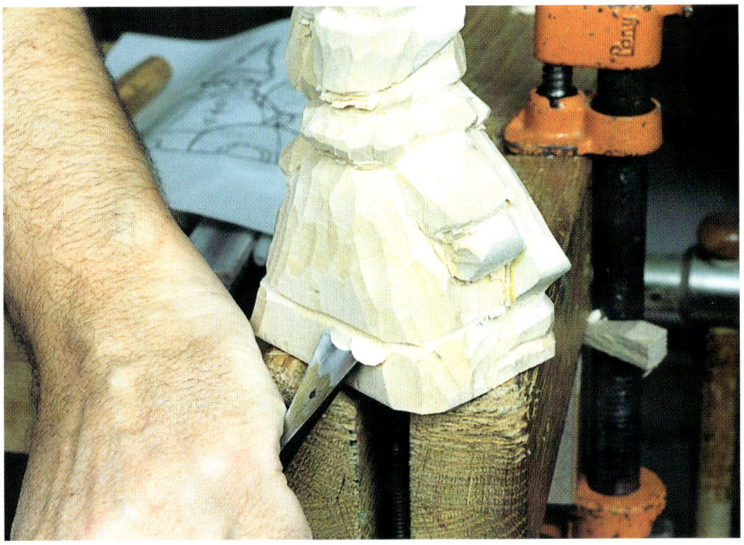

Use the shallow "U" gouge to shape into the stop cut.

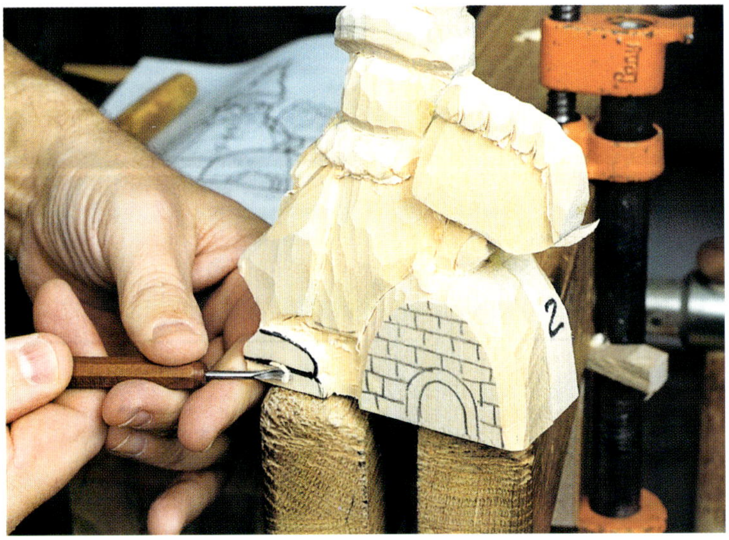

Use a small "V" gouge to cut the line separating the "upper" boot from the "sole."

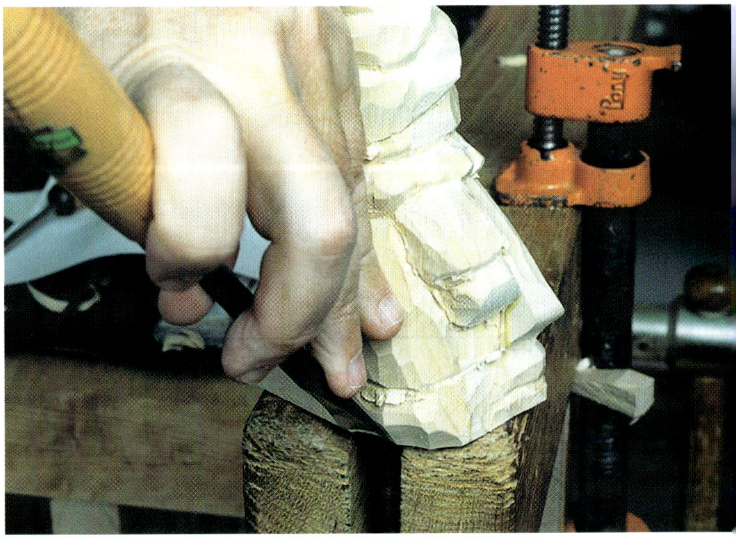

Cut down with the "U" gouge to round off the bottom of the carving.

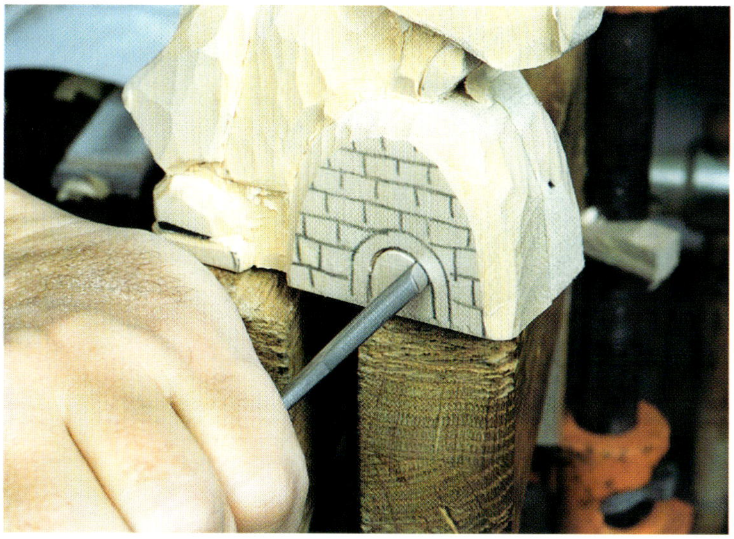

Use a small "U" gouge to outline the line around the igloo entrance.

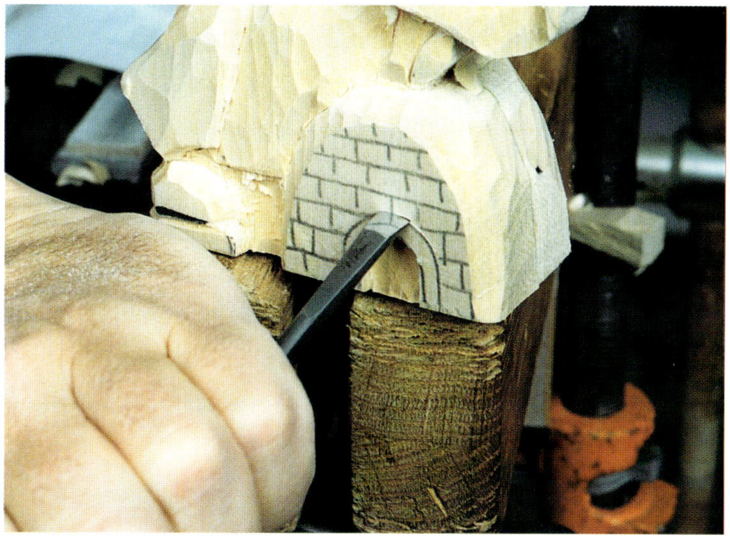

Cut a stop cut around the outside of the igloo entrance.

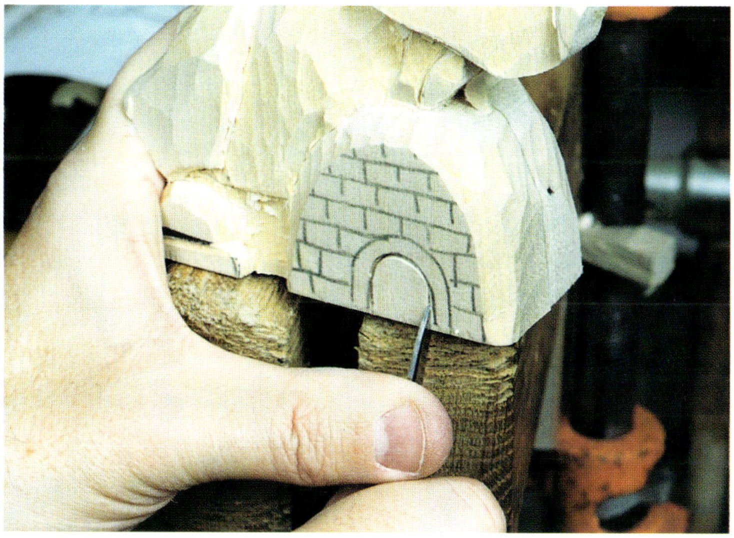

Use a knife to deepen the stop cut.

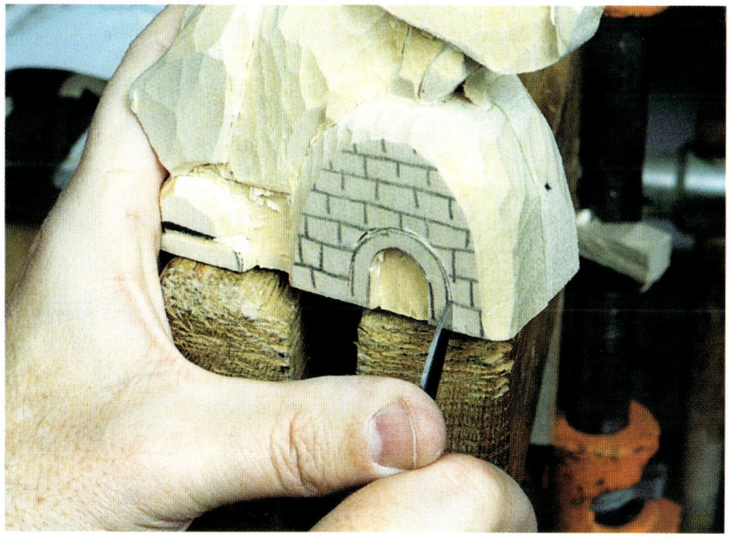

Use the knife to deepen the cut.

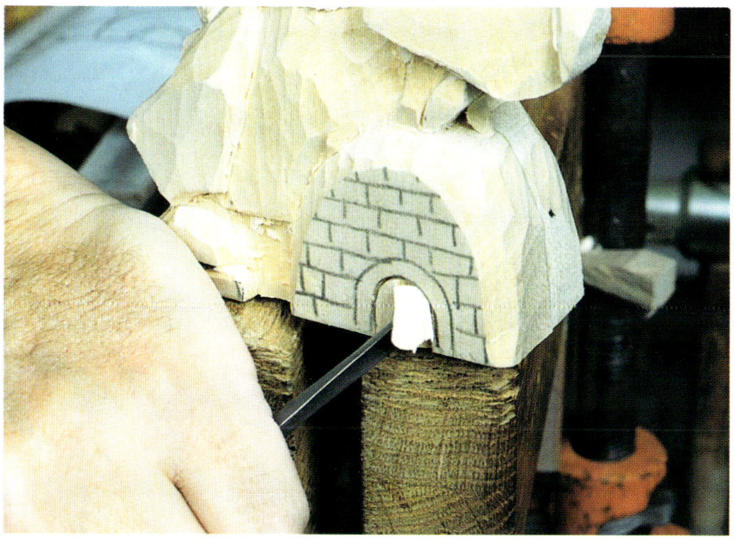

Remove the wood from inside the igloo entrance.

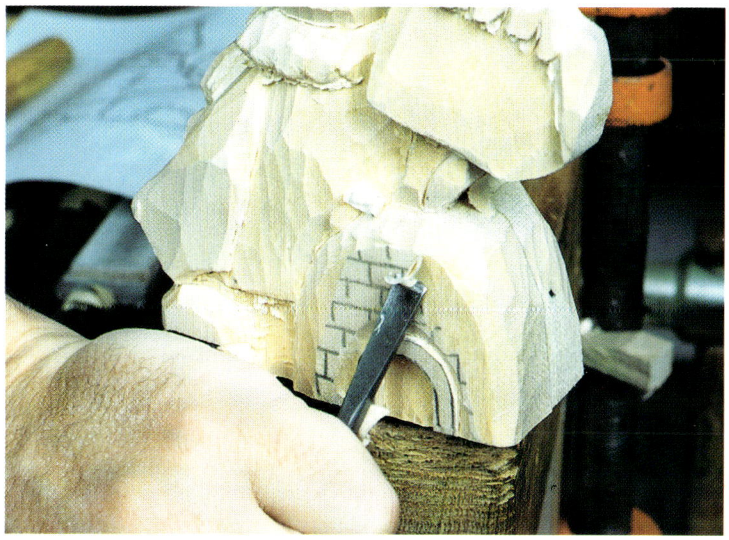

Use a shallow "U" gouge to shape the front and top of the igloo.

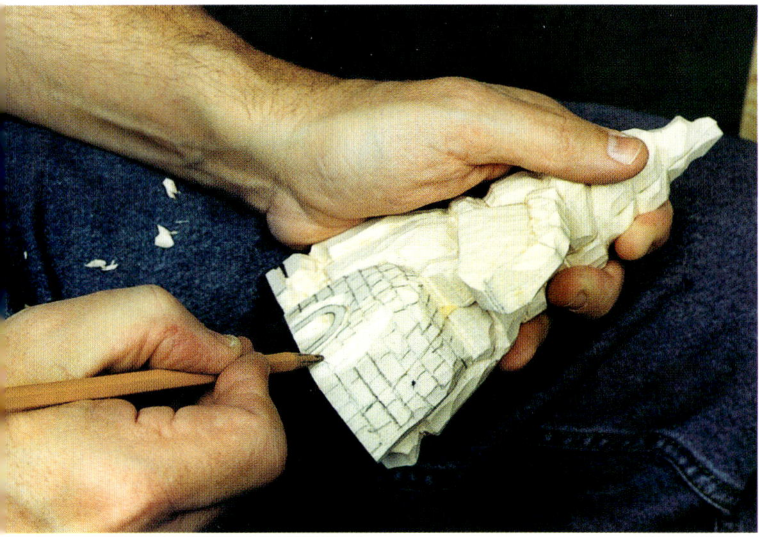

Here I am drawing back the lines that define the ice blocks in the igloo.

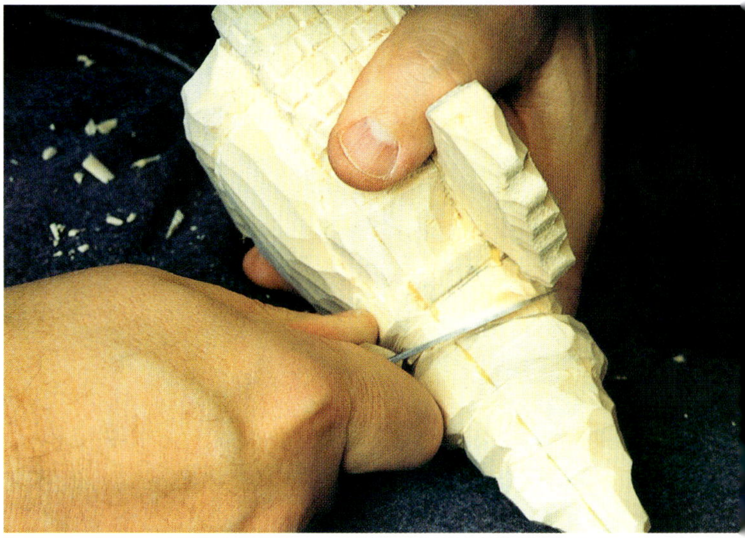

Look the whole carving over and clean up the carving as needed. Cut it back until it pleases you.

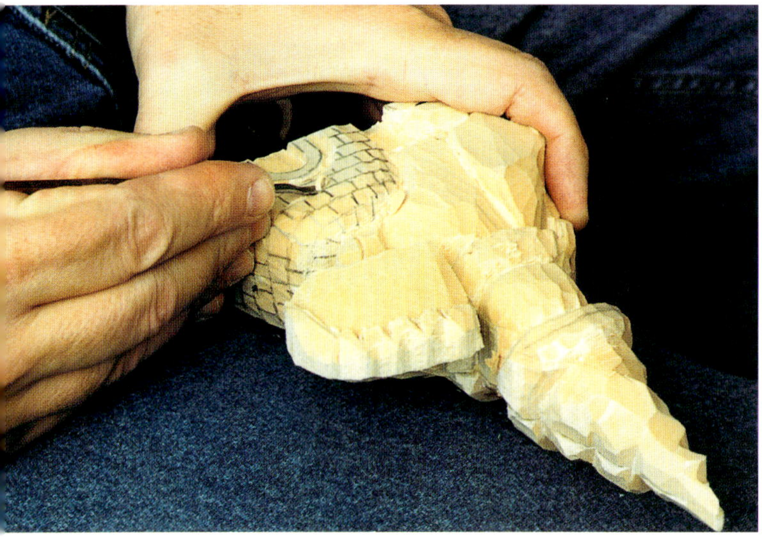

Use a small "V" gouge to cut the lines separating the ice blocks in the igloo.

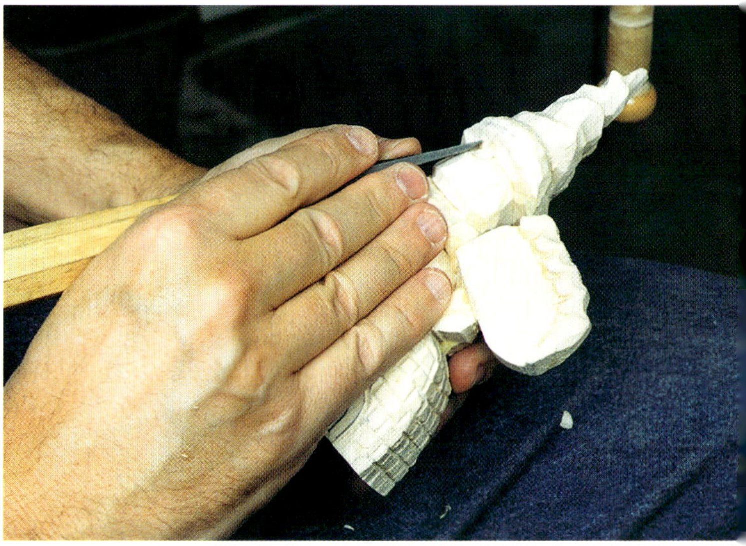

Clean up and define the carving.

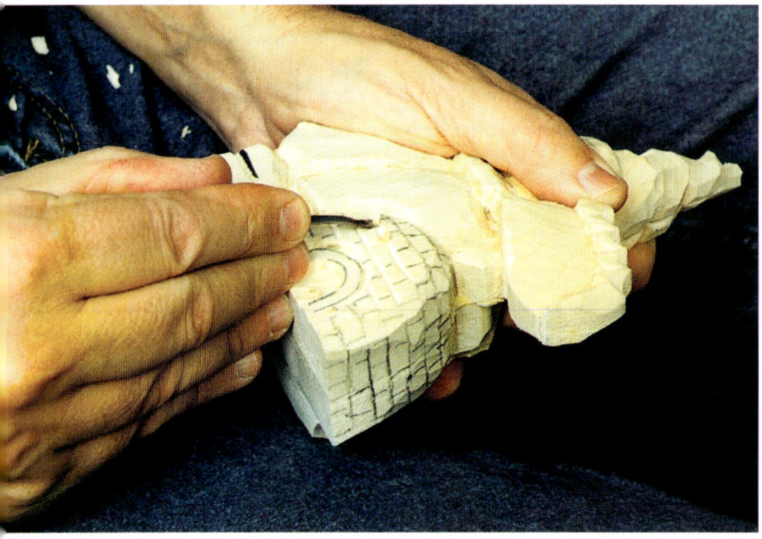

Cut the lines in the igloo.

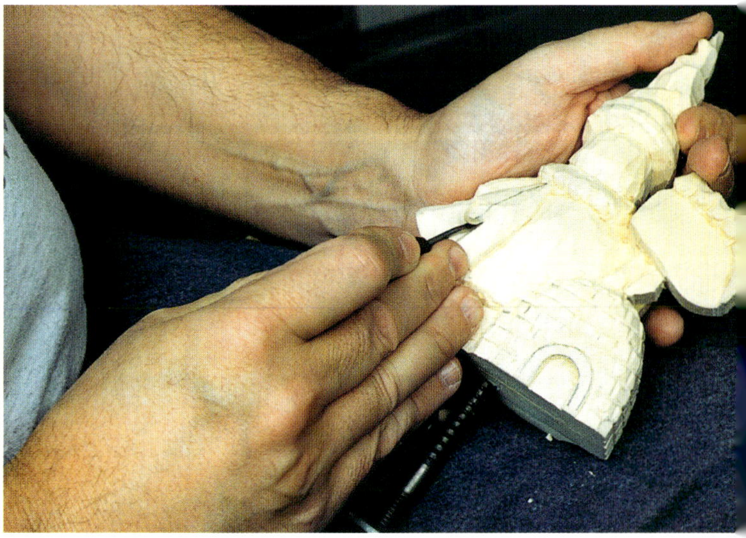

Cut the line separating the two ends of the scarf.

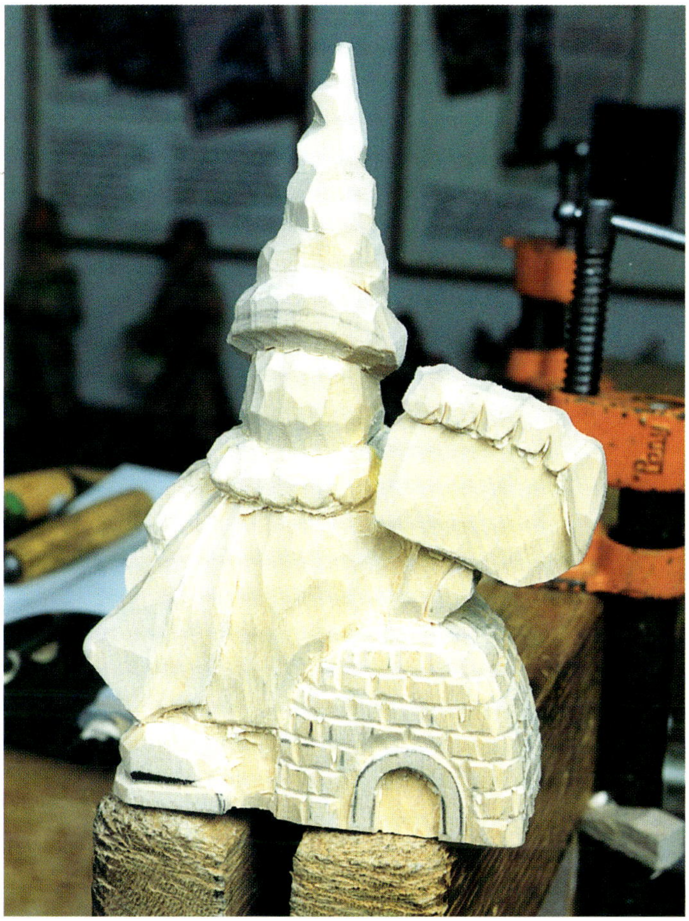

Here is the carving just before I start the sanding process. I use a Foredom rotary tool with ruby and diamond bits to clean up the carving and do some final small shaping and detail work.

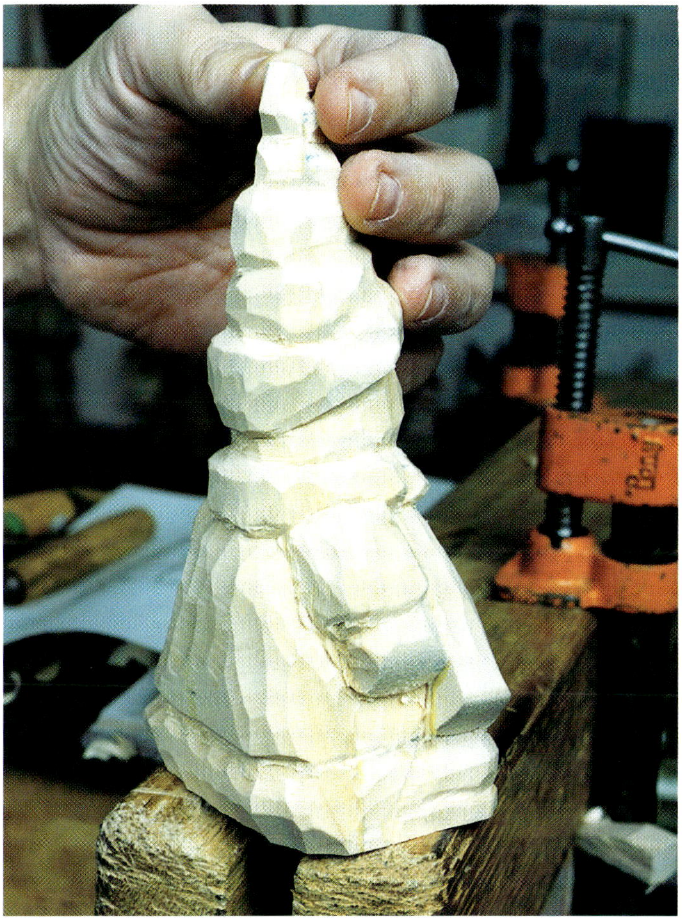

Here is a side view of the Igloo Snowman.

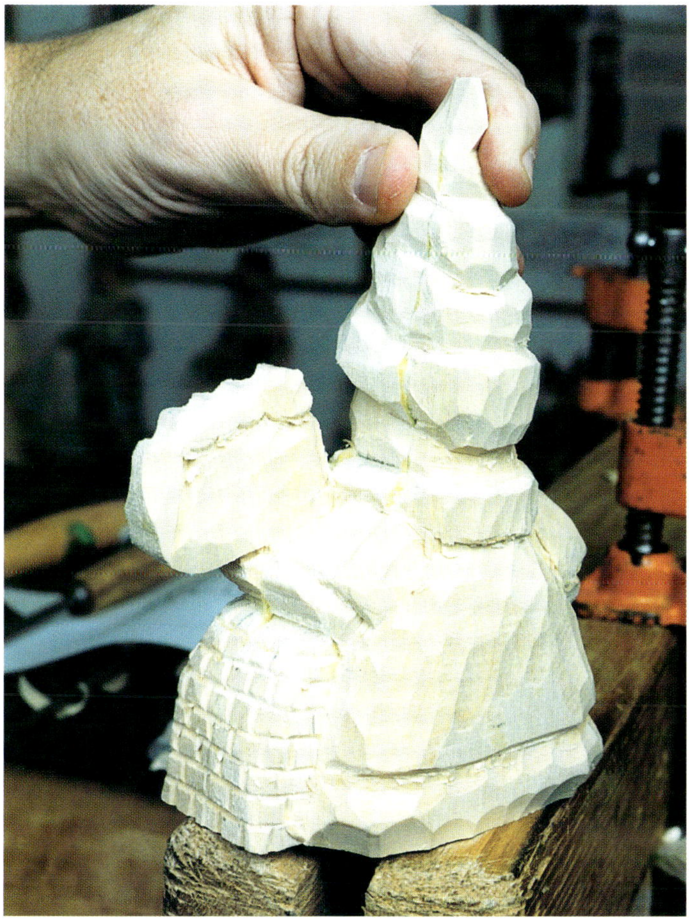

Here is the back view of the Igloo Snowman.

I will sand the piece by hand to make it smooth enough for painting and to give it a "rounder" look appropriate to the idea of a snowman.

Chapter 4

Carving the Snowbank Snowman

Follow the instructions and look closely at the photos to see how I did the Snowbank Snowman.

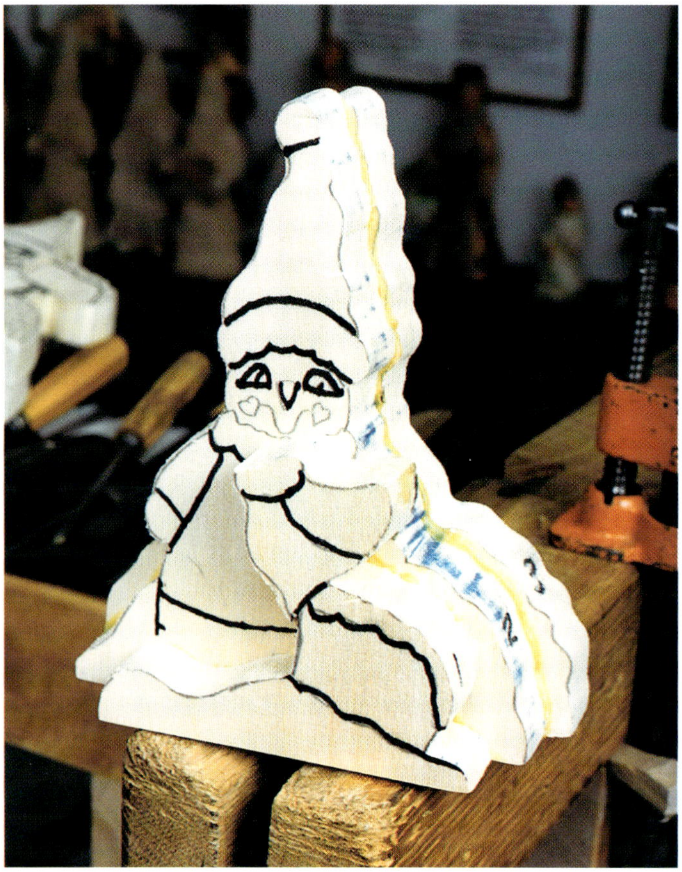

This is the Snowbank Snowman as he has been glued up from the three layers. I highlighted him with dark ink lines so you can see where we will be working.

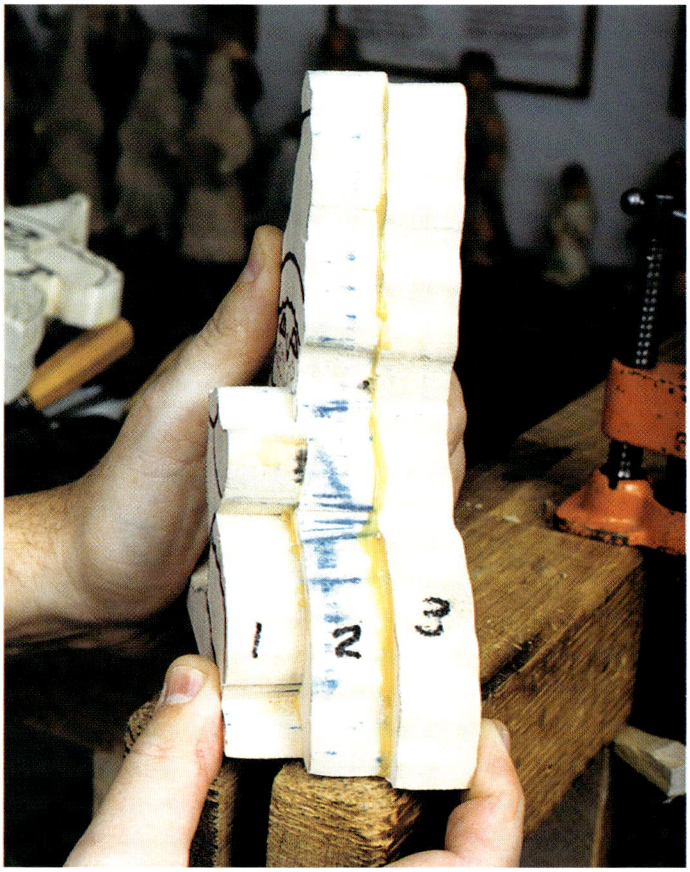

This is the side view of the Snowbank Snowman. The three layers are numbered from front to back.

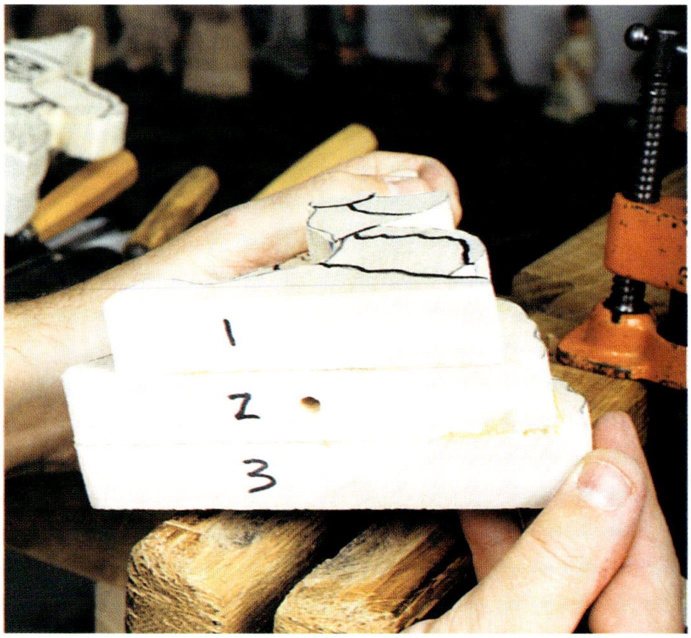

This is the bottom of the Snowbank Snowman showing the pilot hole I drilled with a 1/4" drill. This hole will accept the 3/8" carver's screw.

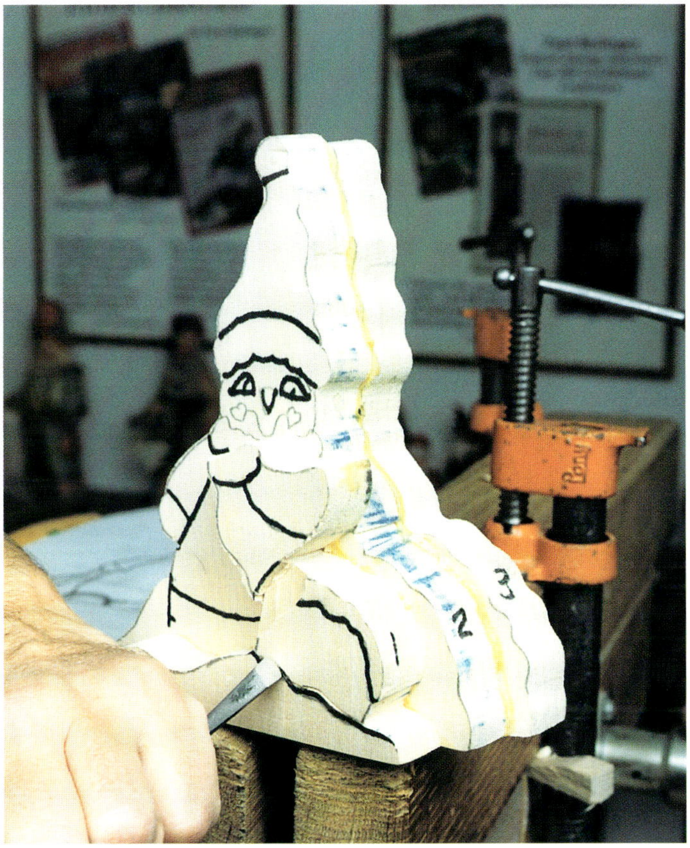

Cut a stop cut into the snowbank along the bottom of the sign. Cut down to the line, removing some of the sign, then cut up to remove some of the sign face.

Cut on the snowbank to round it somewhat.

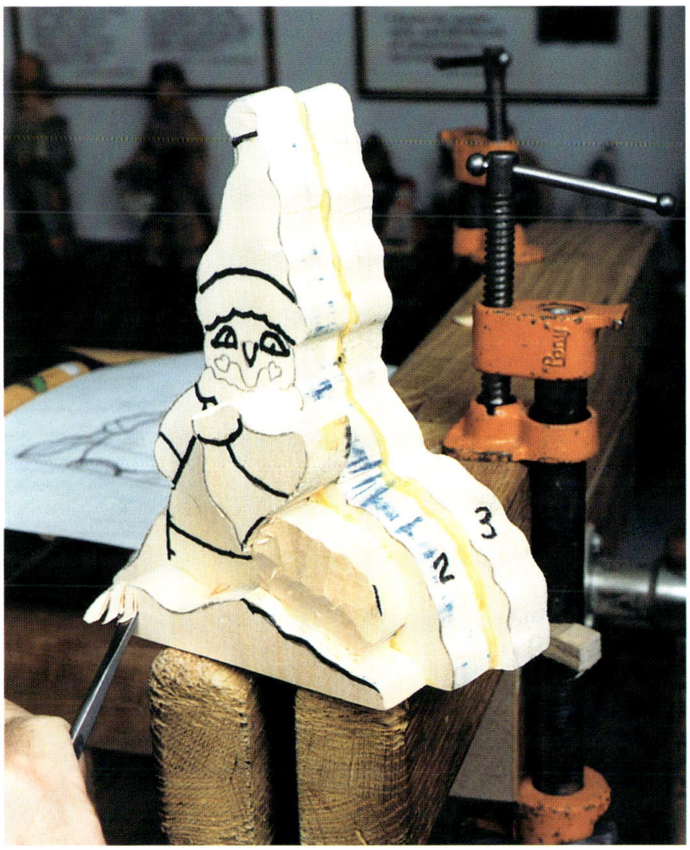

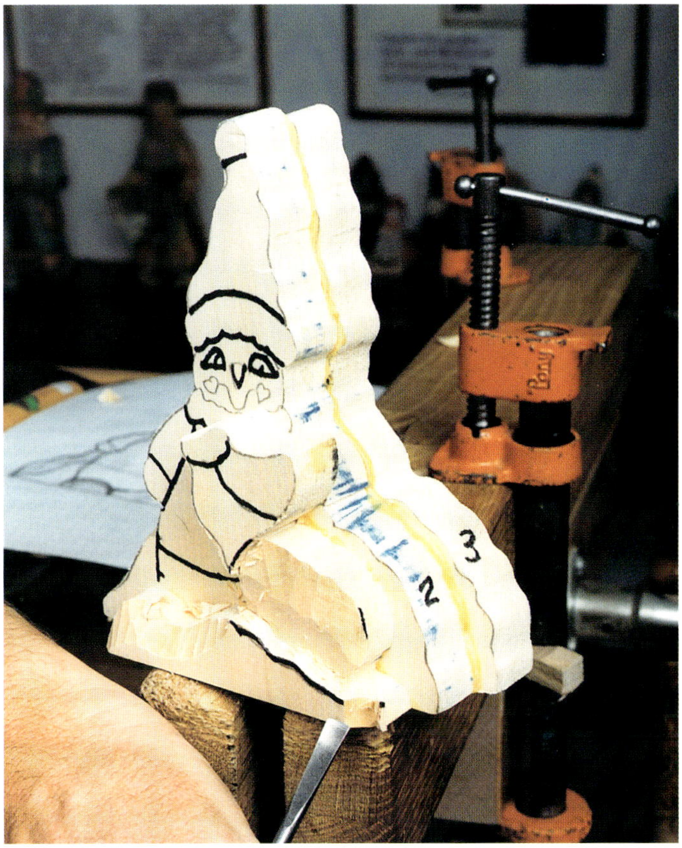

Cut on the snowbank's right side to round off the snowbank.

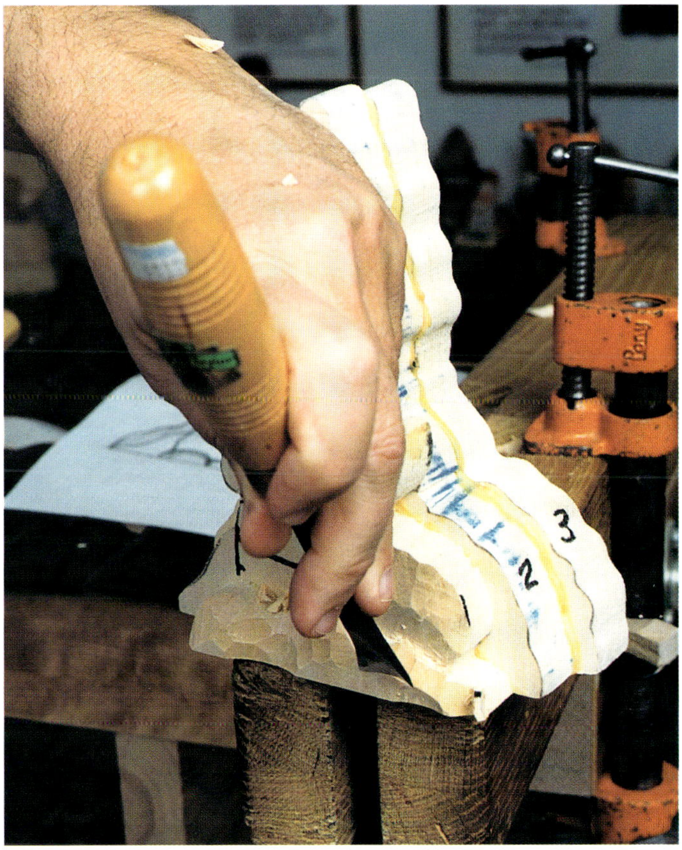

Cut down at the bottom of the snowbank to make it look more like a snowbank.

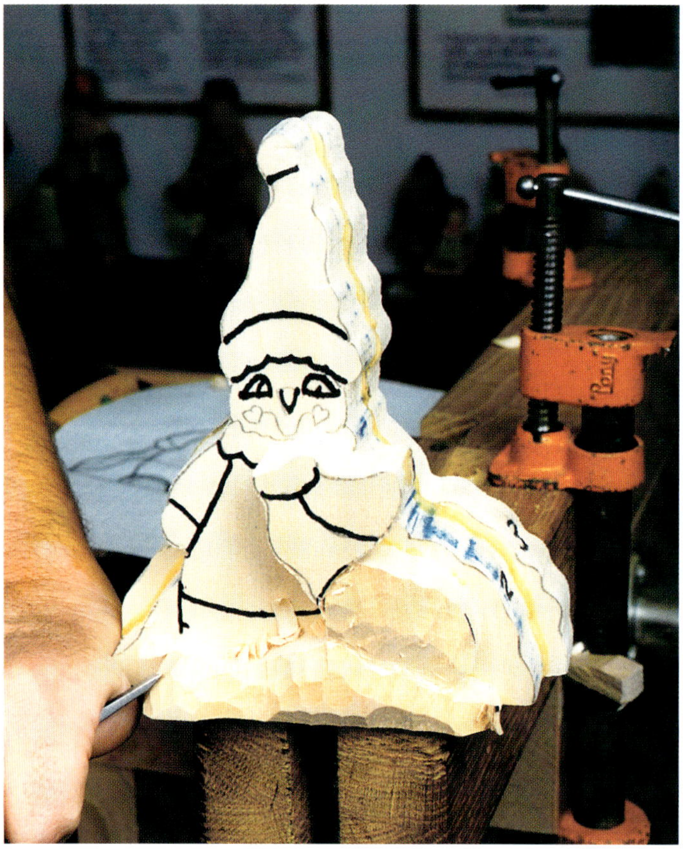

Work on the snowbank.

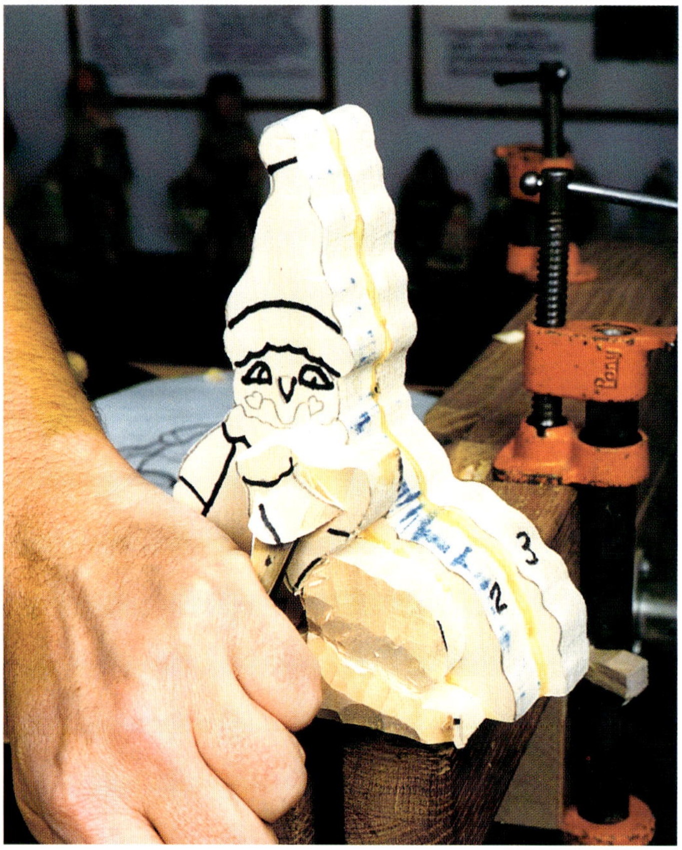

Use a large shallow "U" gouge to cut the scarf back at an angle from the bottom up to the chin of the snowman.

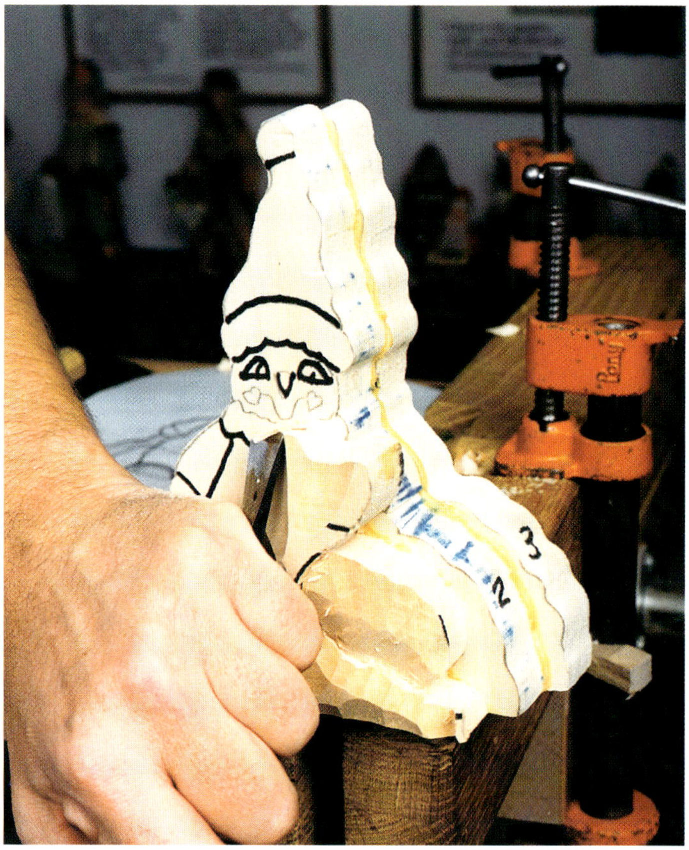

Cut the scarf to a shape that flows from the top of the sign to the chin.

Continue to shape the scarf.

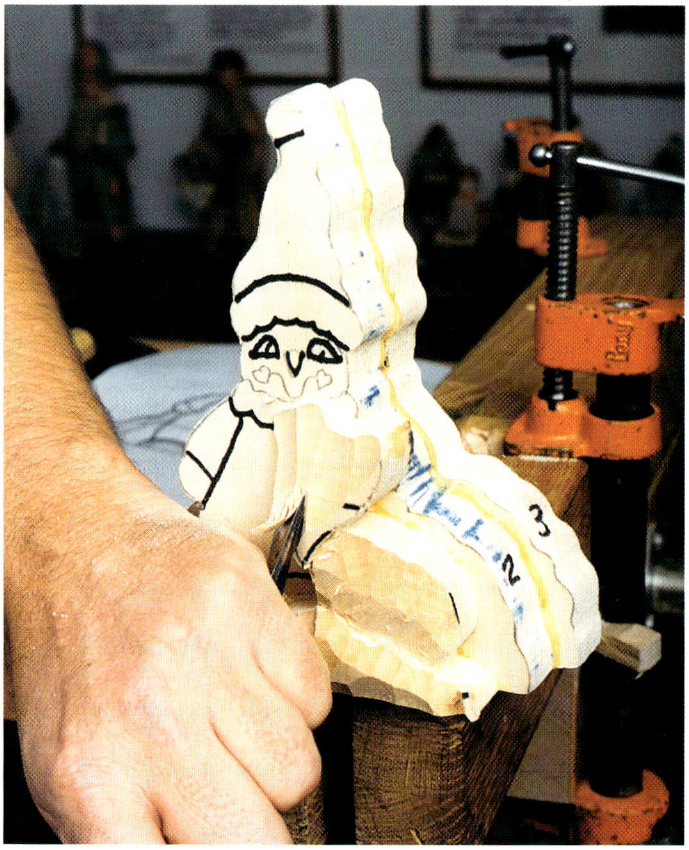

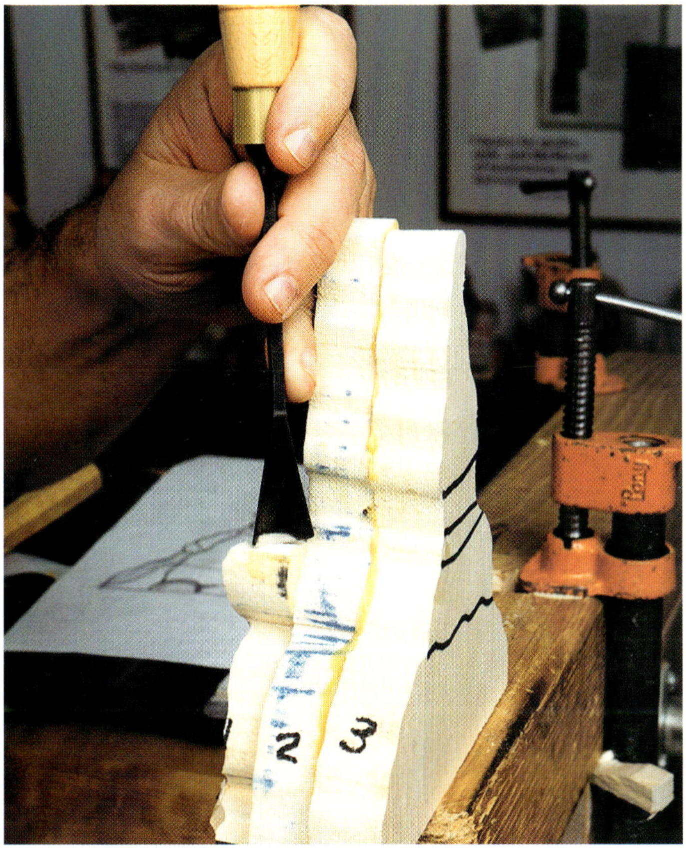

Use a large gouge to cut away the back of the scarf, thinning it down so it looks more cloth-like. Be careful not to break off the scarf – it is somewhat delicate.

Work on the Snowman's right side – see the numbered layers. Shape the snowbank so it looks smoother, like a snowbank.

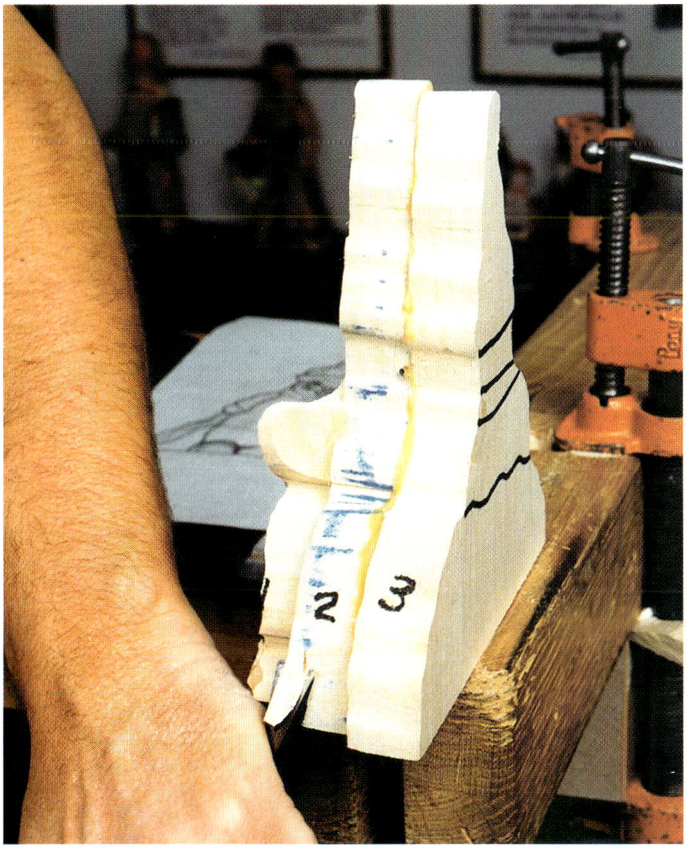

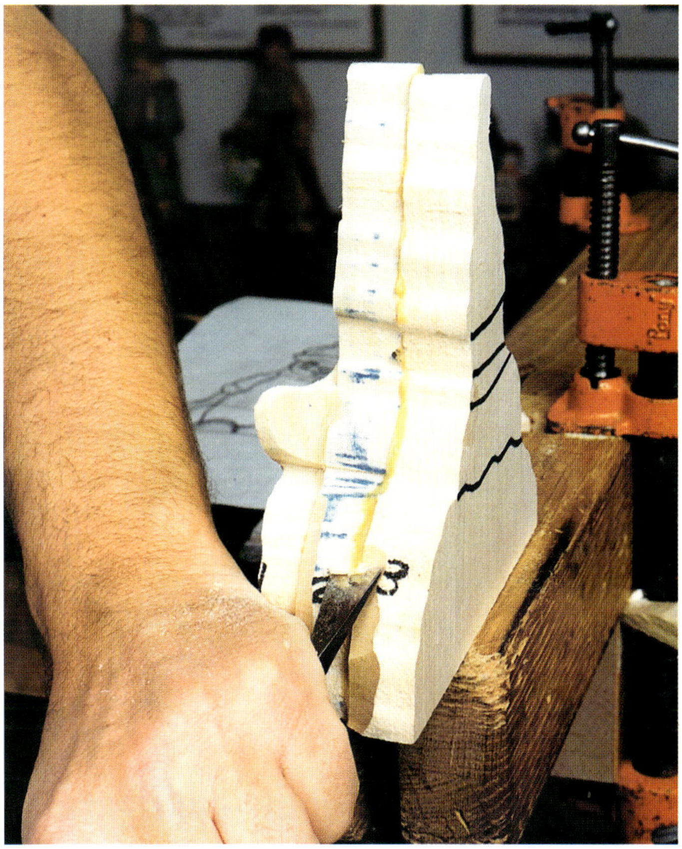

Shape the right side of the snowbank.

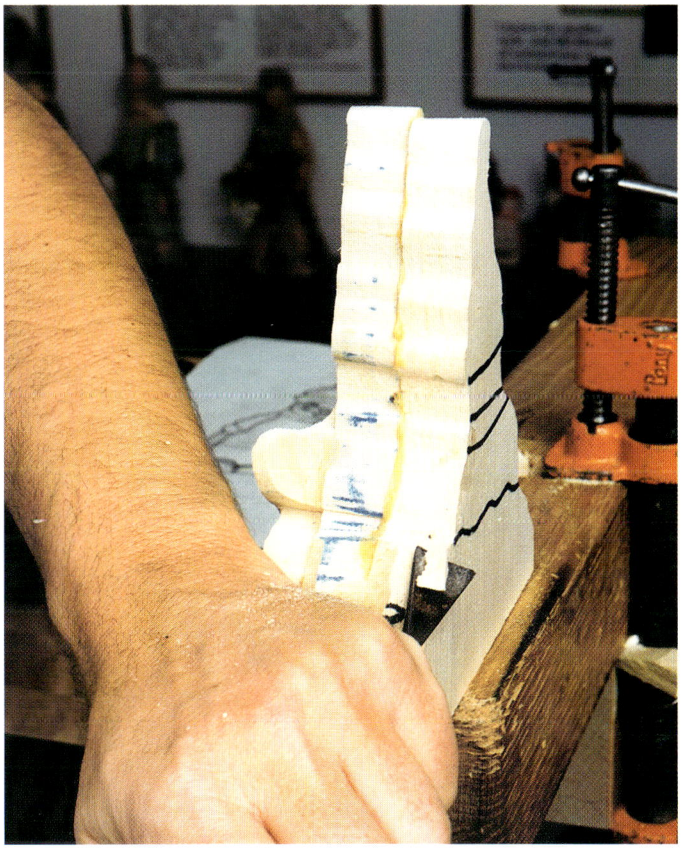

Shape the side where it meets the back to start the rounding of the snowbank.

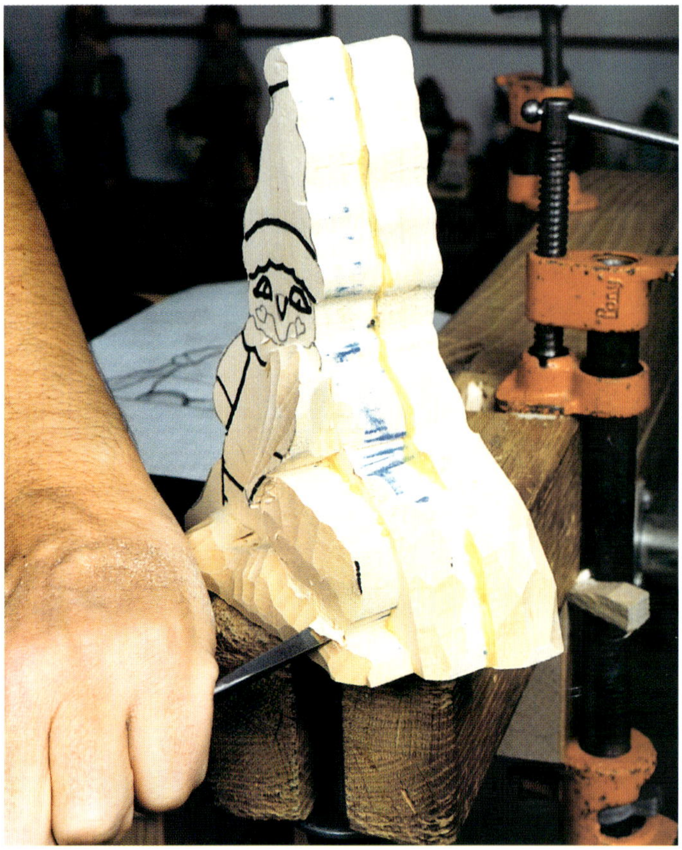

Return to the snowbank under the sign. Remember that we keep moving around to avoid over-cutting any one spot.

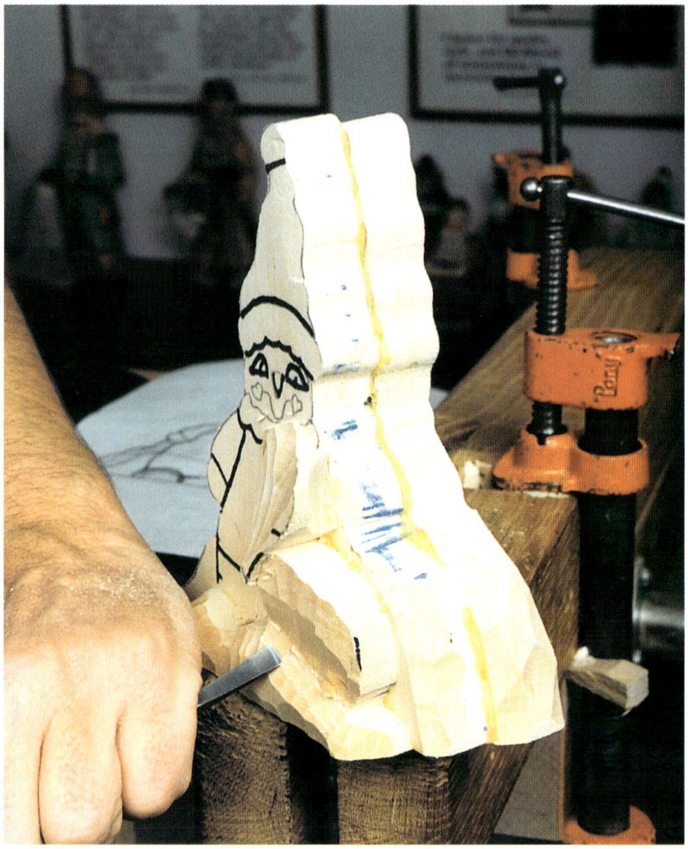

Cut up at the top of the snowbank.

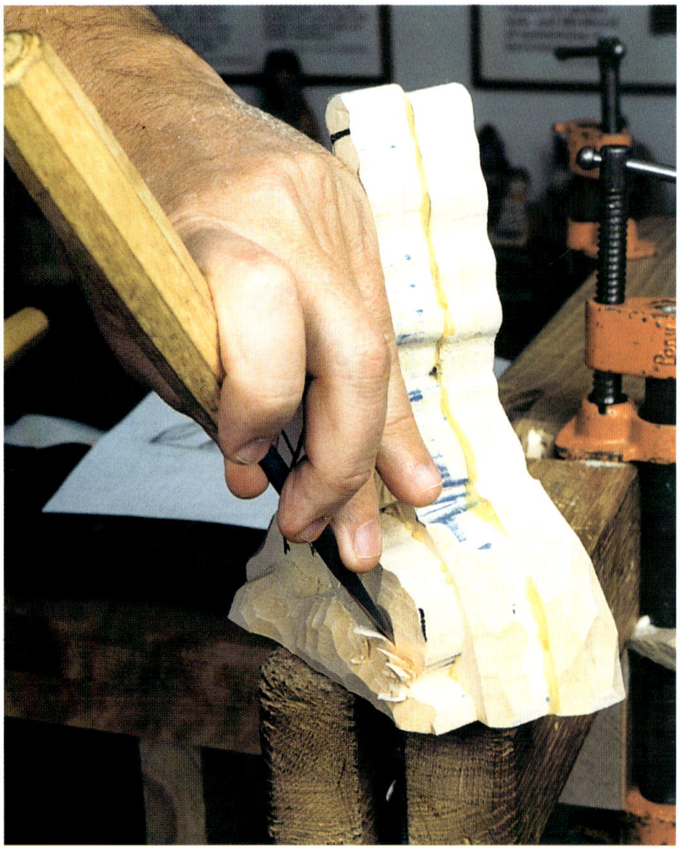

Cut down at the bottom of the sign to further define the sign.

Make a stop cut at the bottom of the scarf knot.

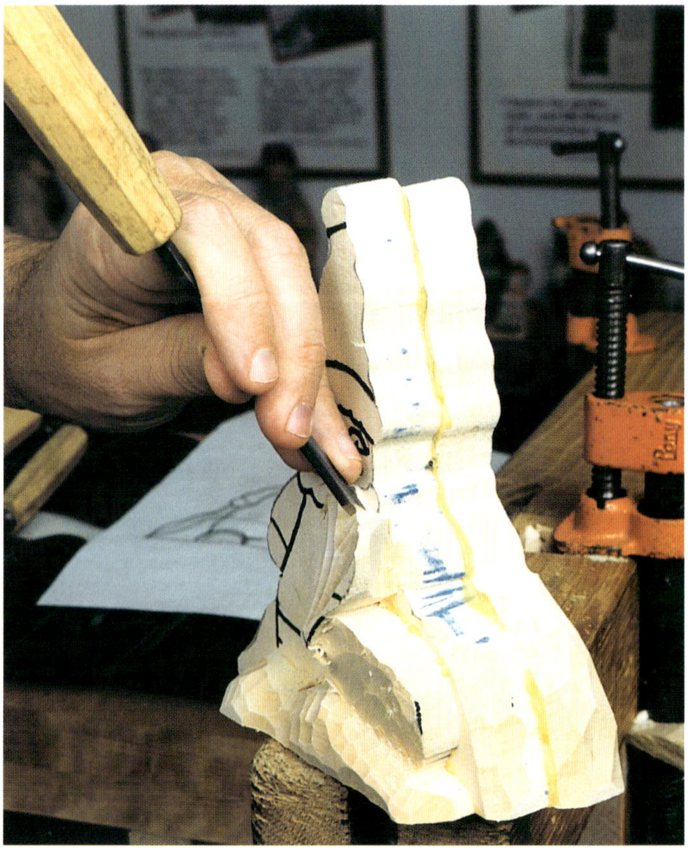

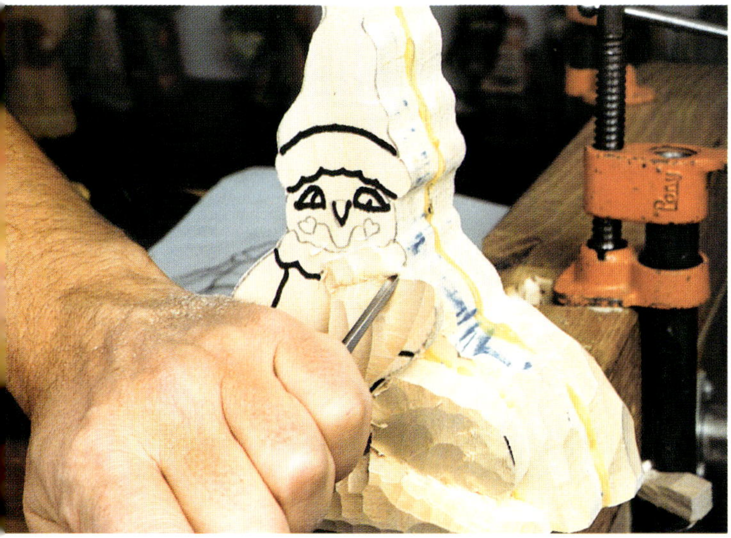

Cut up to the stop cut and define the bottom of the knot on the scarf.

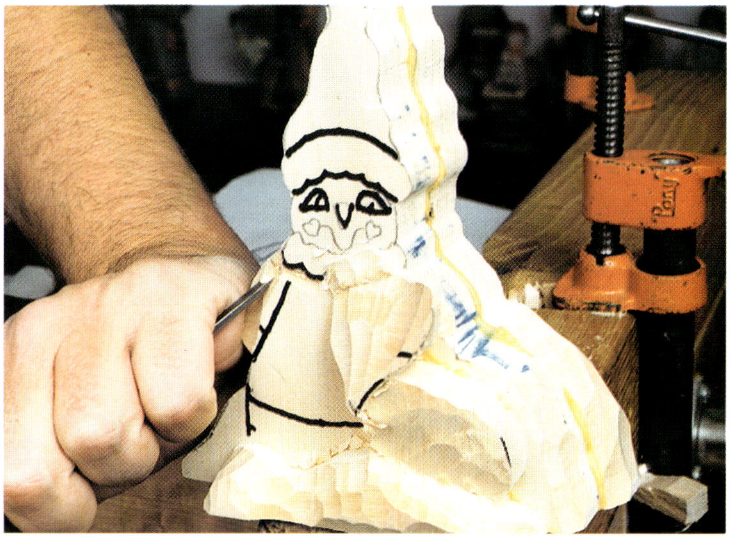

Start to shape the left side of the Snowman.

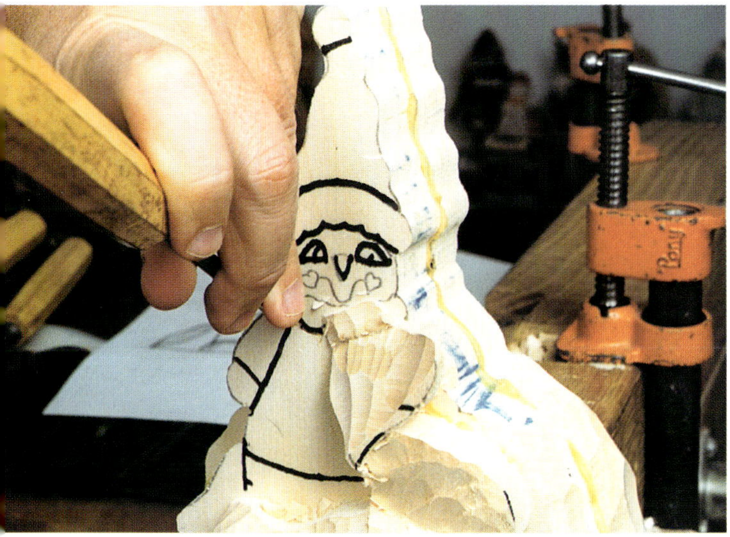

Cut a stop cut at the bottom of the scarf.

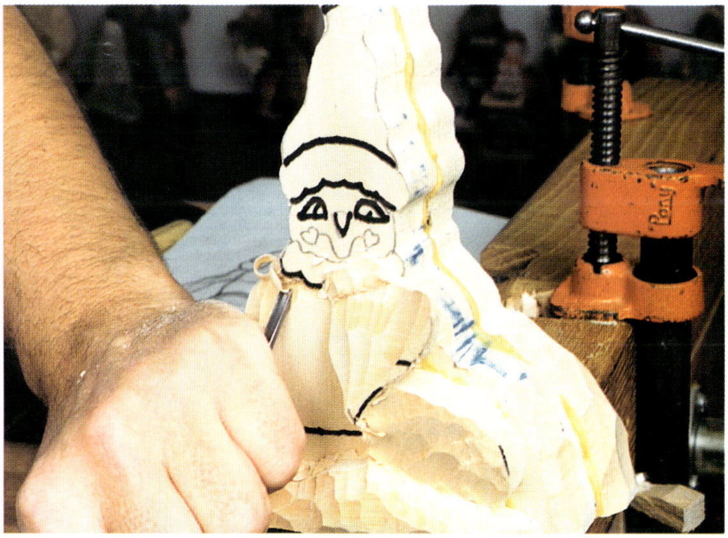

Shape the left front side of the snowman.

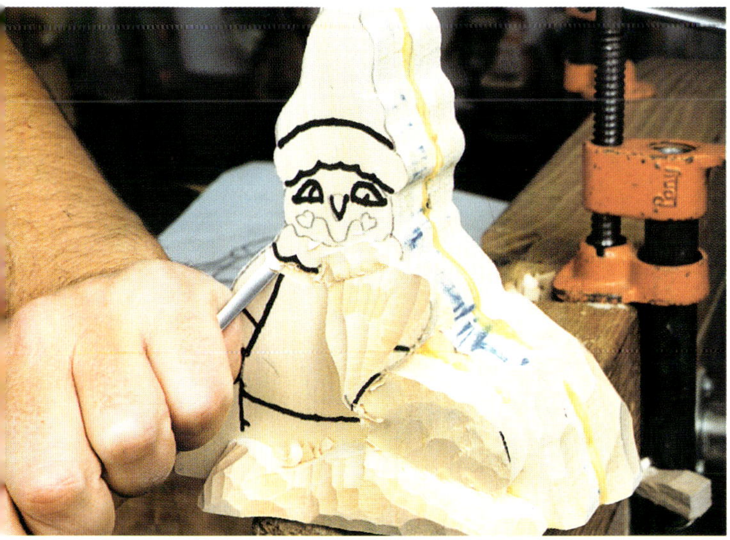

Cut up to the bottom of the scarf. Remove some of the snowman below the scarf to start the process.

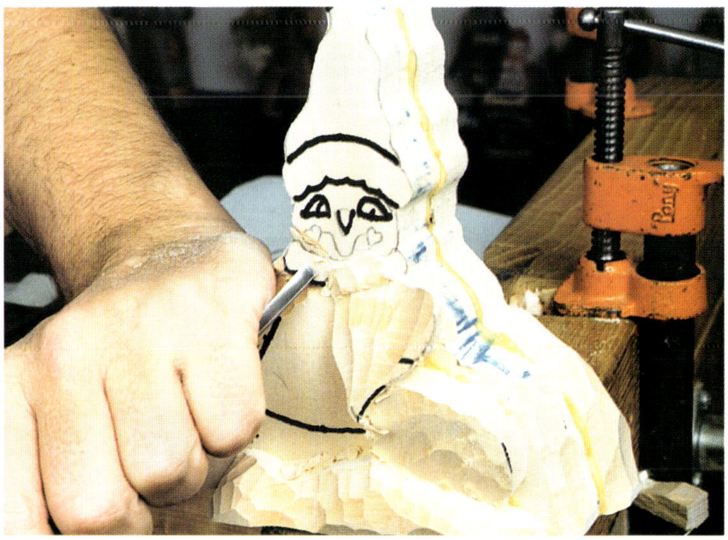

Make a stop cut along the top of the scarf and then cut down to the stop cut as well as up to the stop cut. This will begin to define the area at the bottom of the Snowman's face.

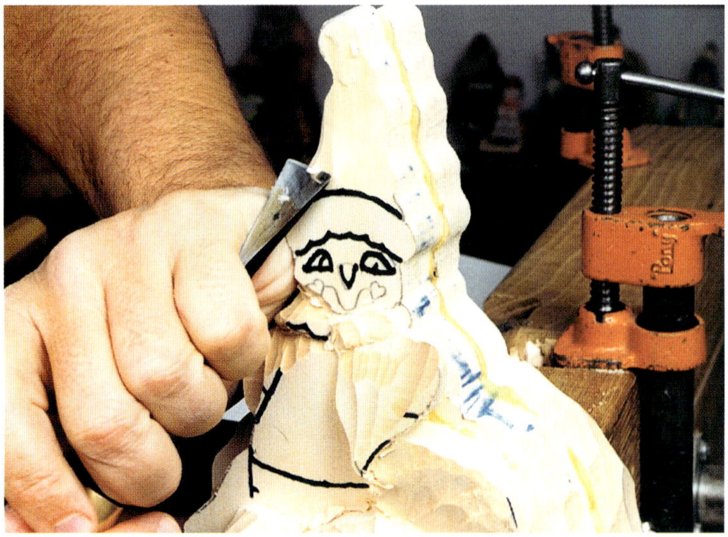

Use a large shallow "U" gouge to start shaping back the hat. The hat will lean back so that the face will be easier to see.

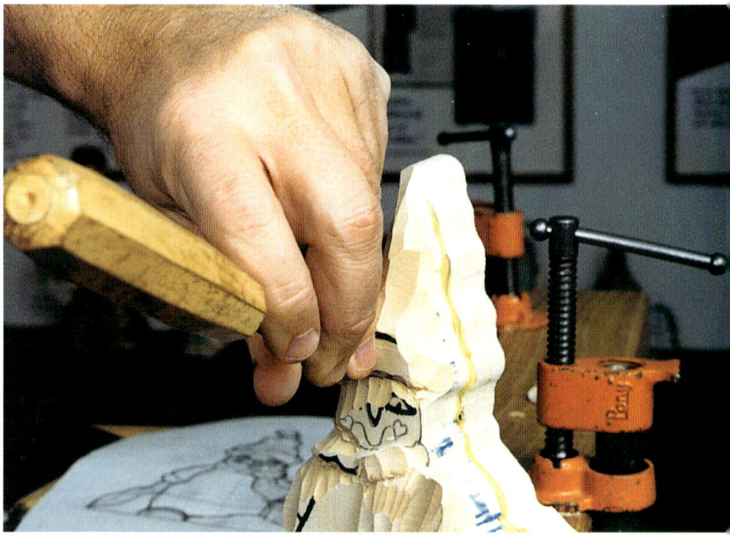

Deepen the stop cut at the bottom of the hat.

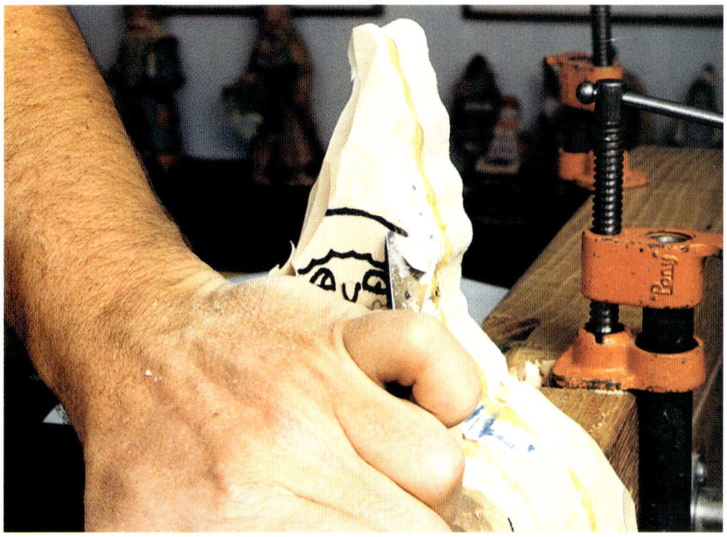

Shape the hat.

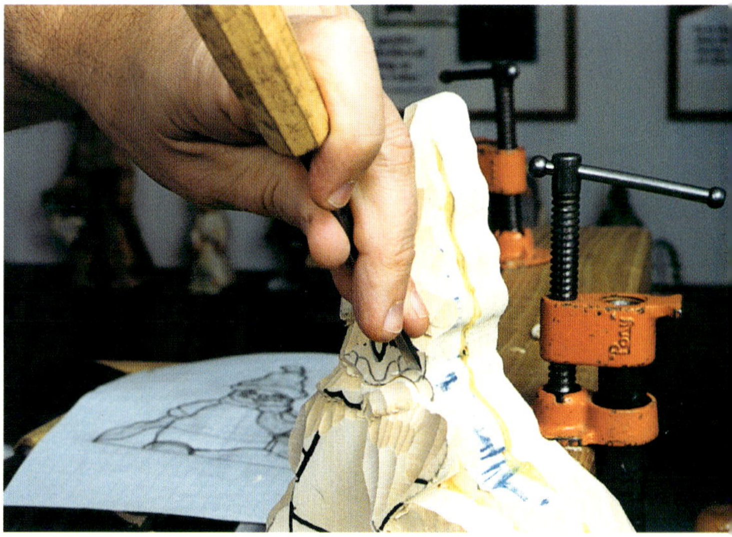

Make a stop cut along the top of the scarf moving around to the side of the piece.

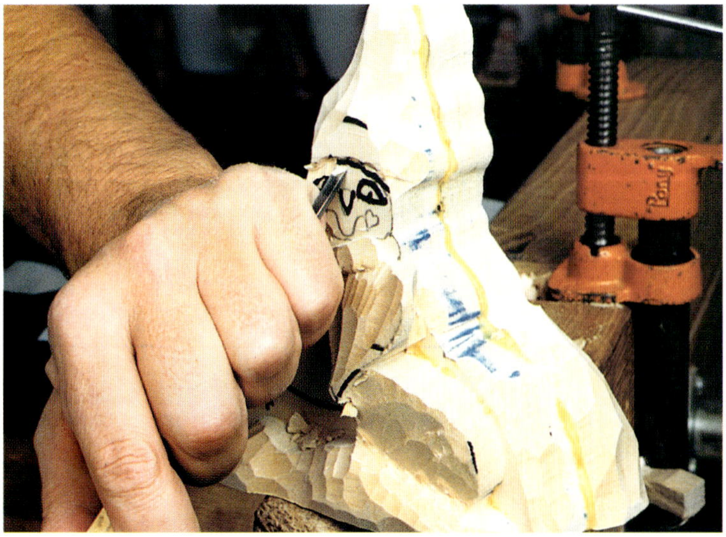

Cut a stop cut along the bottom of the hat. Begin cutting up to the stop cut to start shaping the upper face of the Snowman.

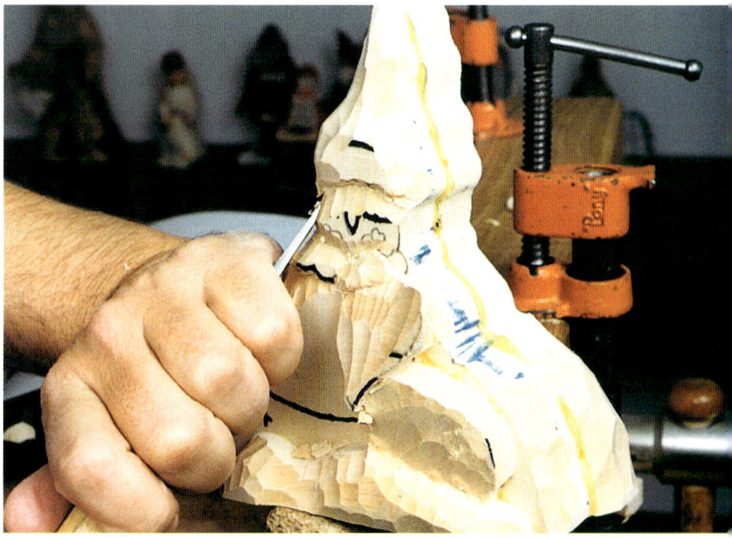

Shape up to the stop cut.

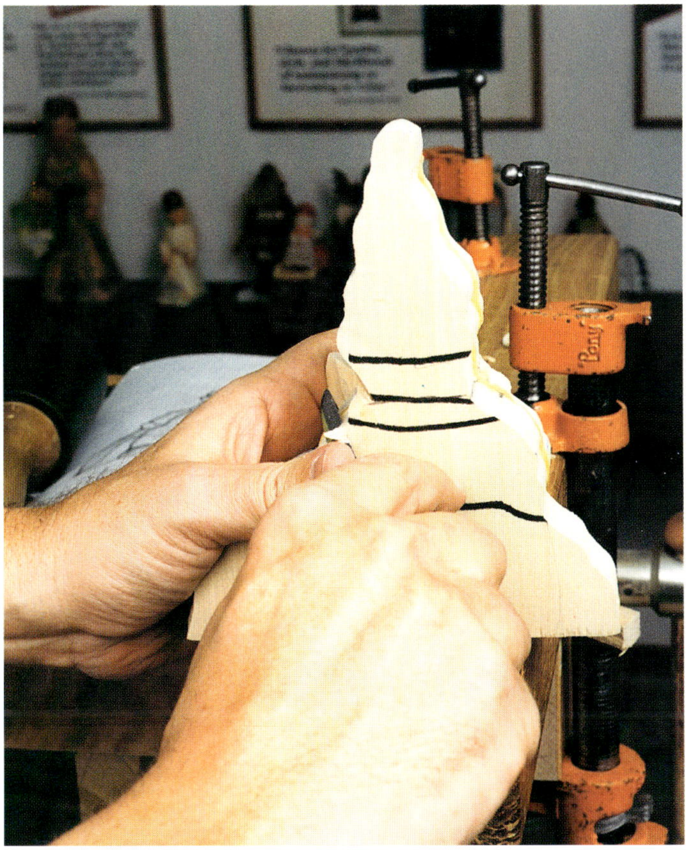

Use a knife to begin shaping the back of the piece. The major areas are the hat, the scarf, the upper body, and then the snowbank at the bottom.

Use a knife or large gouge to start shaping the back of the hat.

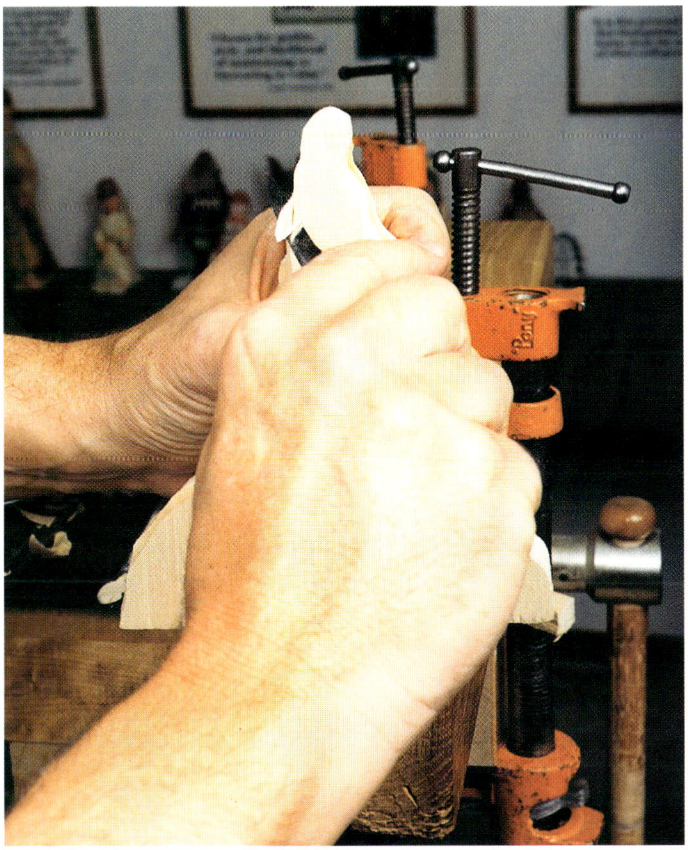

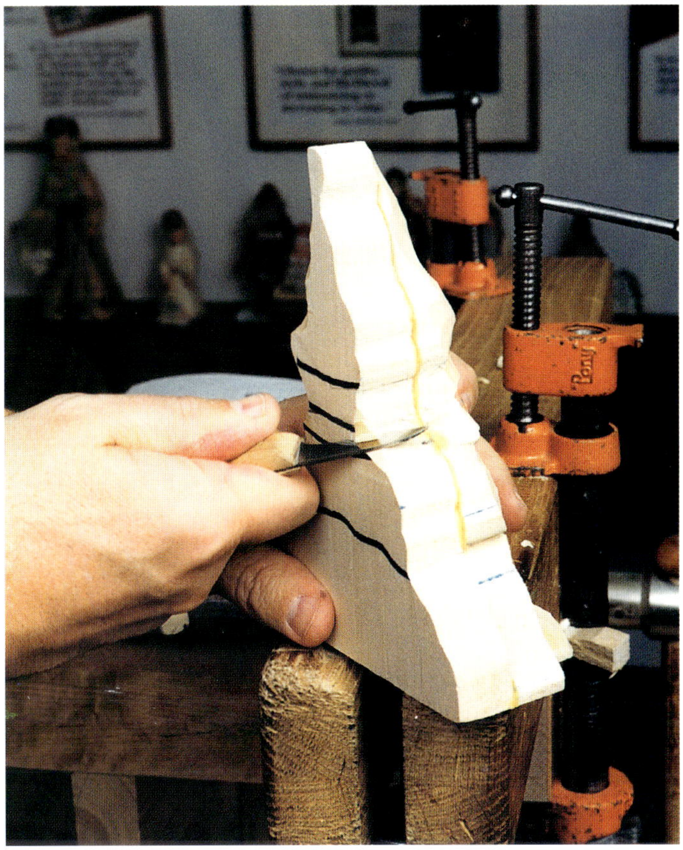

Begin shaping the other side of the upper portion of the piece.

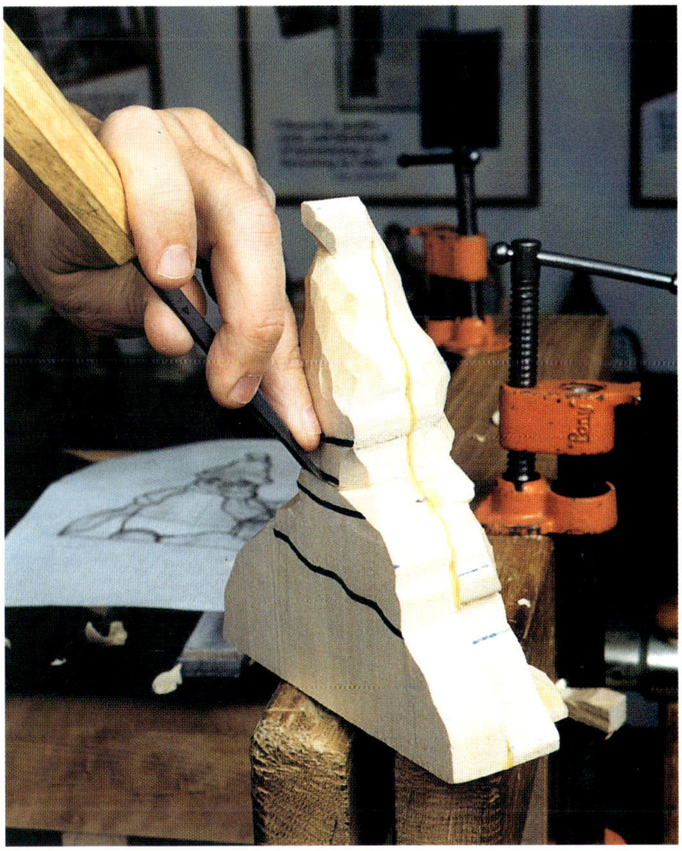

Make stop cuts along the lines at the back. These stop cuts will define the hat and the scarf.

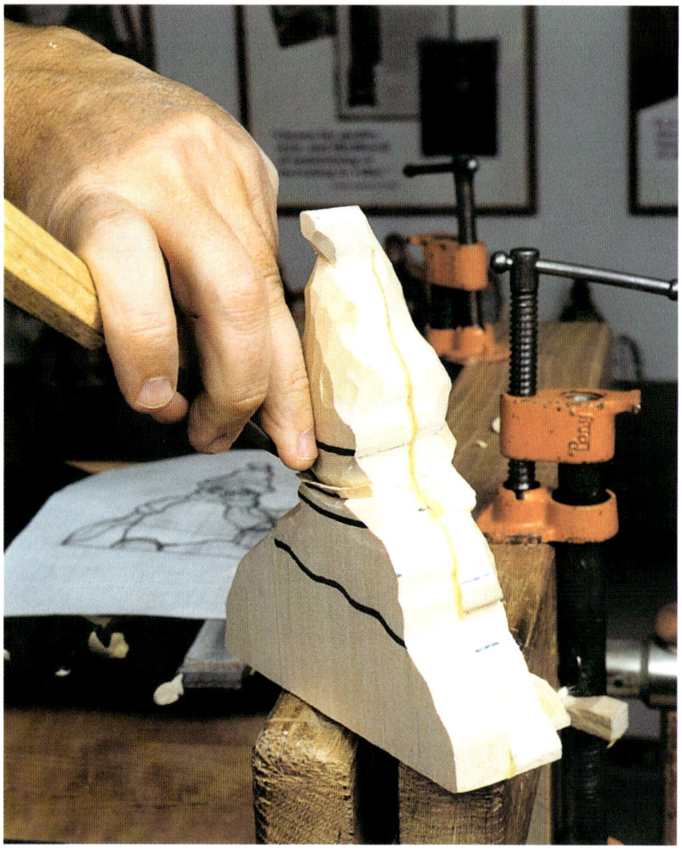

Cut down into the stop cut.

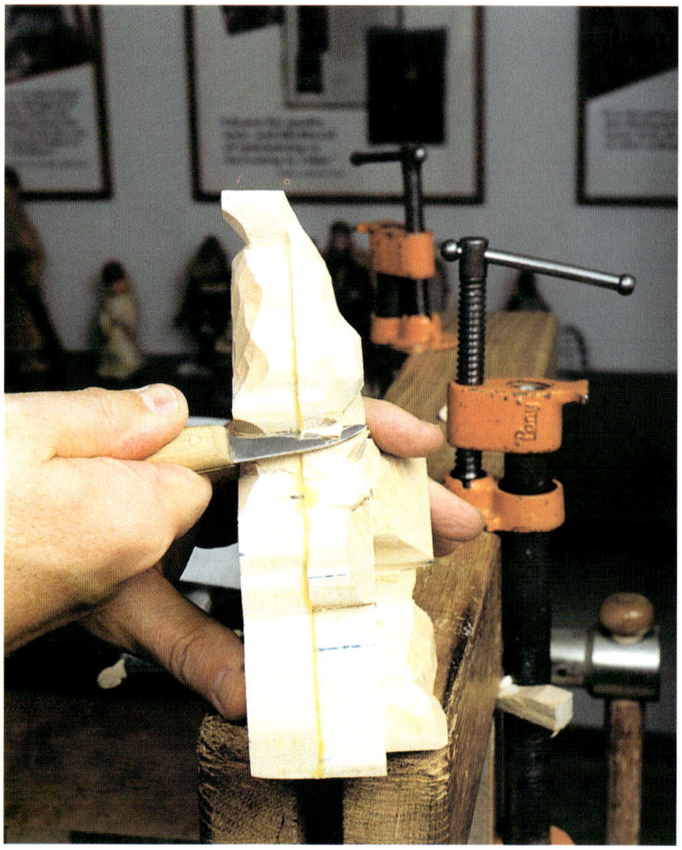

Use the knife to define the cuts on the side of the piece.

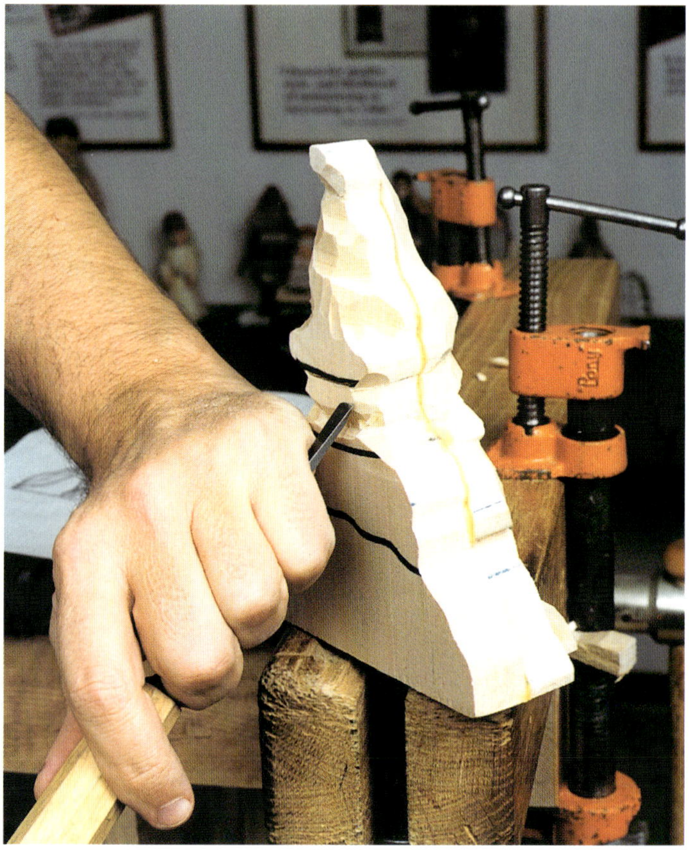

Cut up into the stop cuts along the back to begin defining the piece at the back.

Cut down to the top of the scarf.

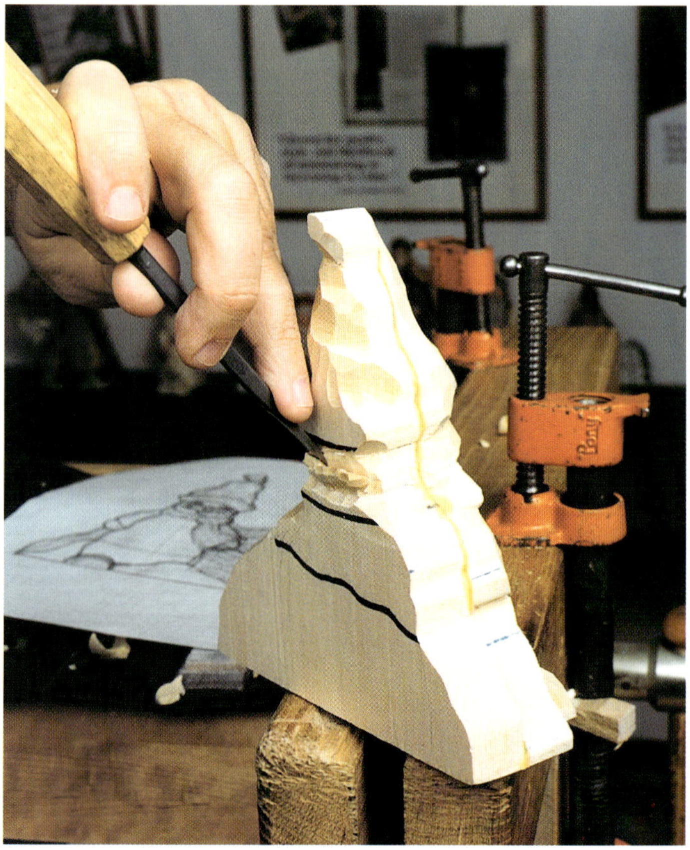

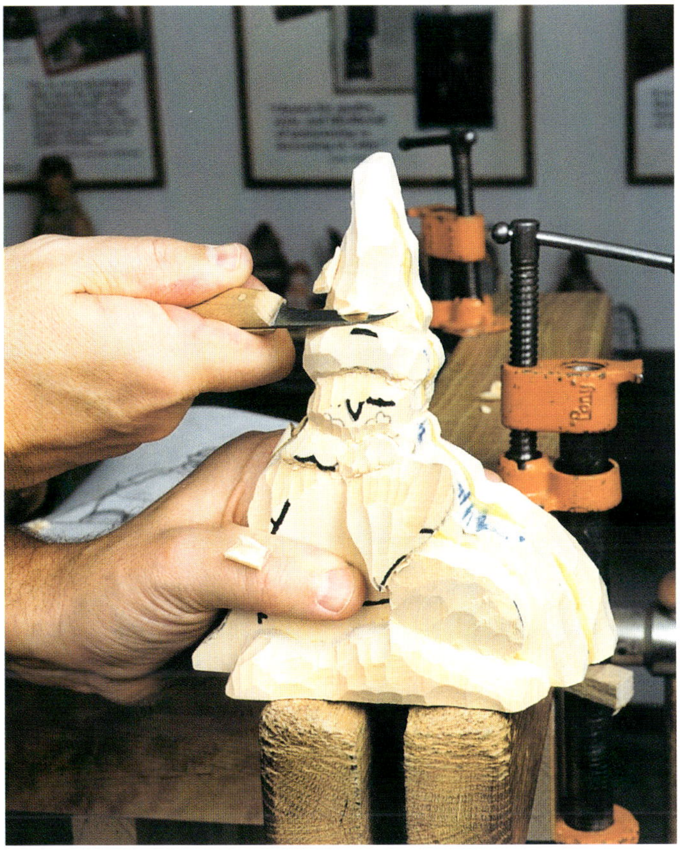

Start shaping the front of the hat. The actual shape is up to you. You may use the basic form or make more dramatic cuts as you wish.

Shape the hat.

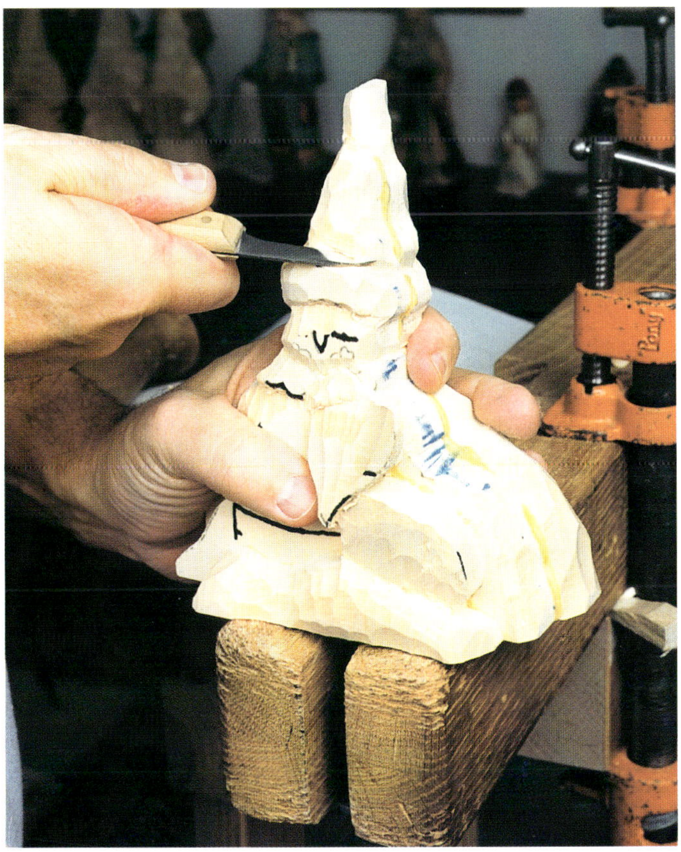

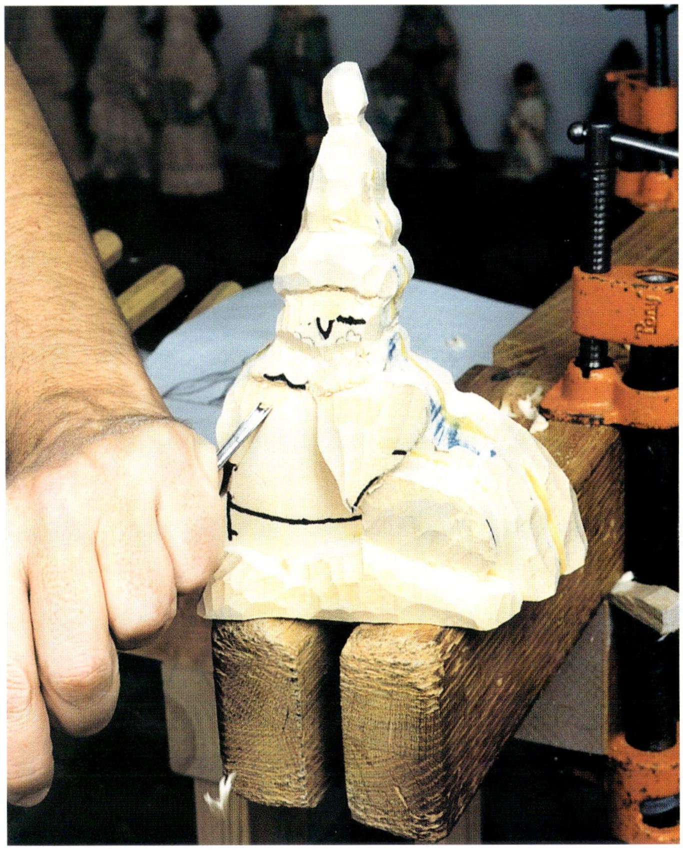

Shape the upper torso cutting up to the bottom of the scarf.

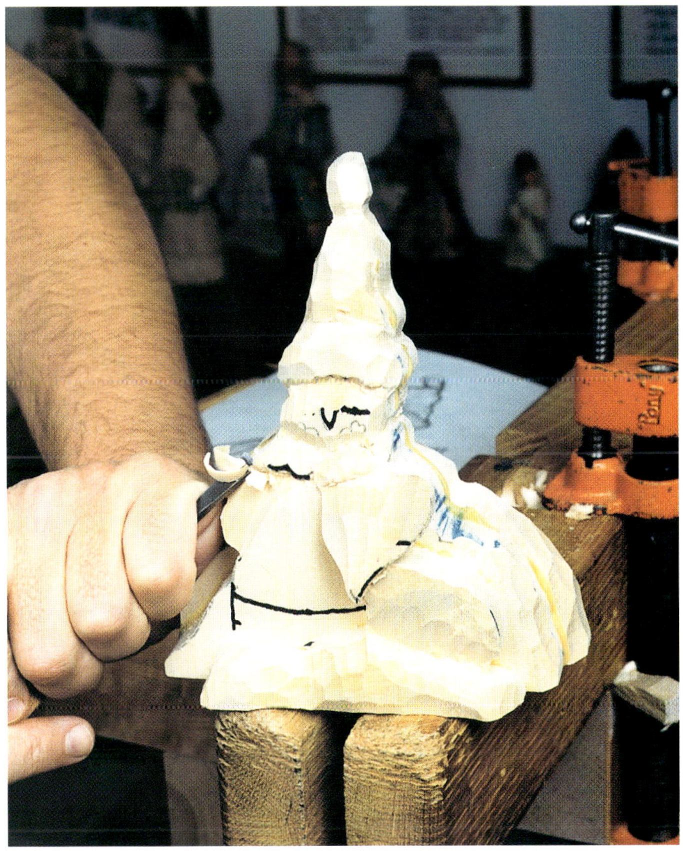

Shape the arm, at the side, up to the bottom of the scarf.

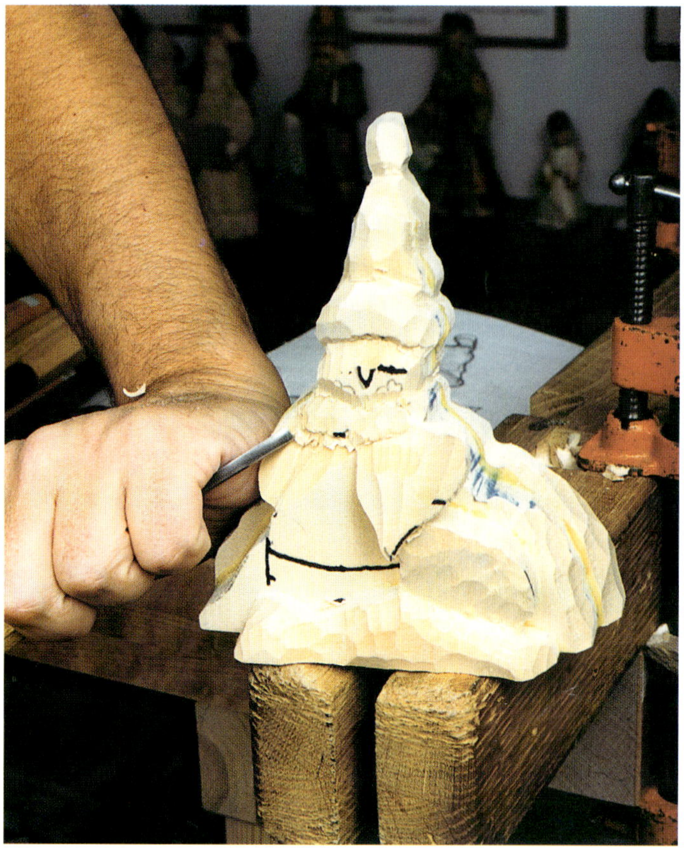

Shape the side and left front.

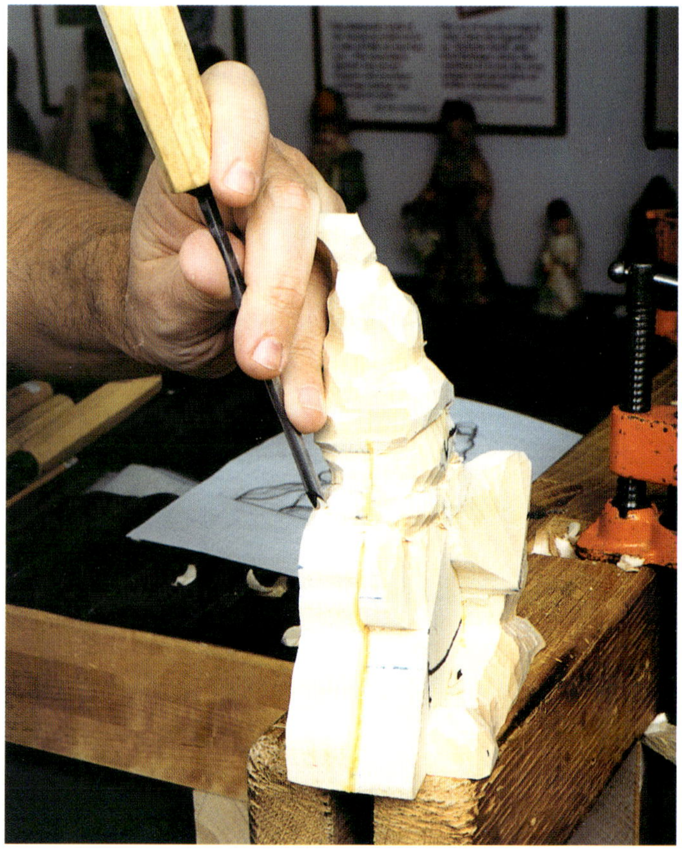

Make a stop cut along the bottom of the scarf.

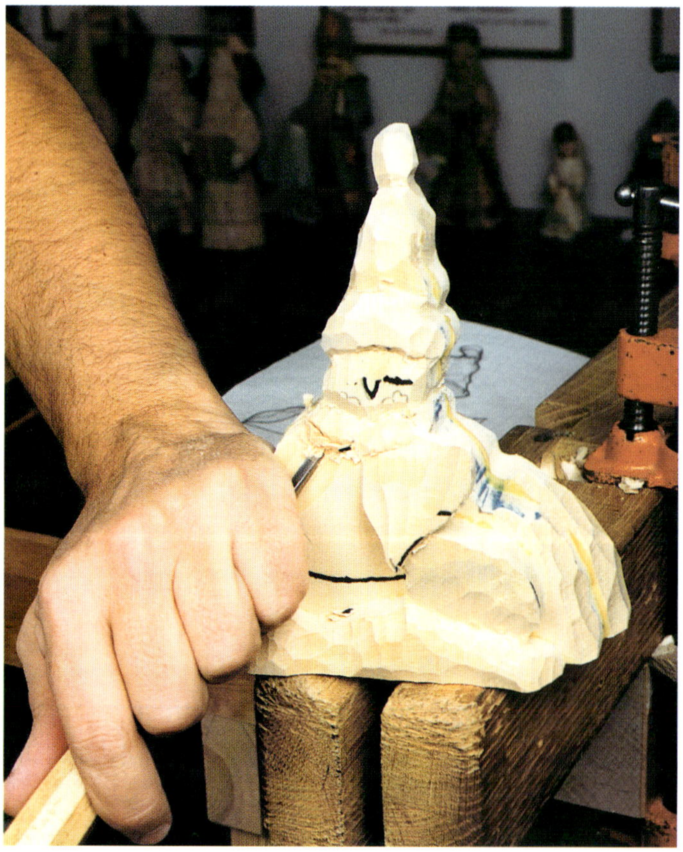

Deepen the area below the scarf.

Cut up under the scarf.

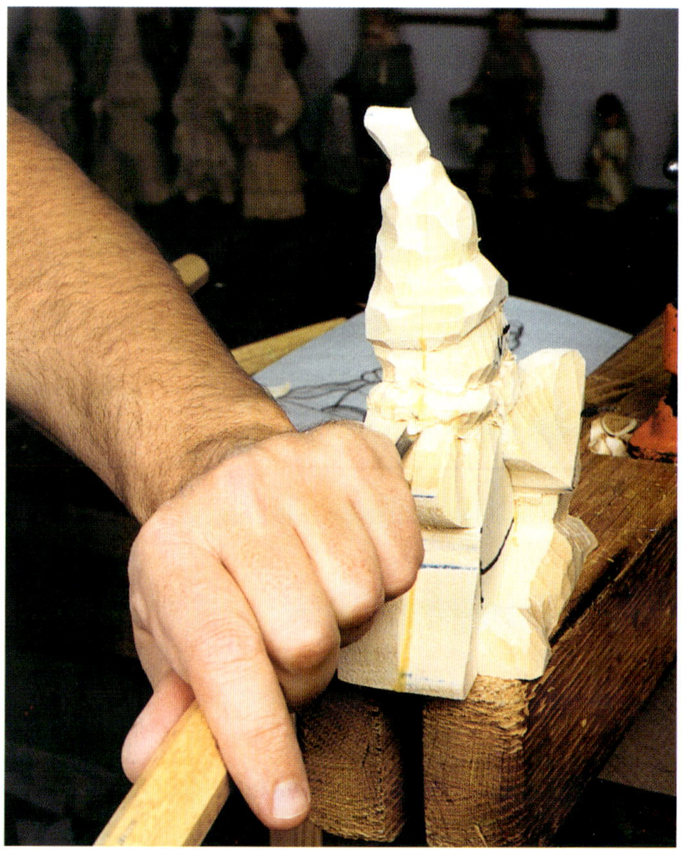

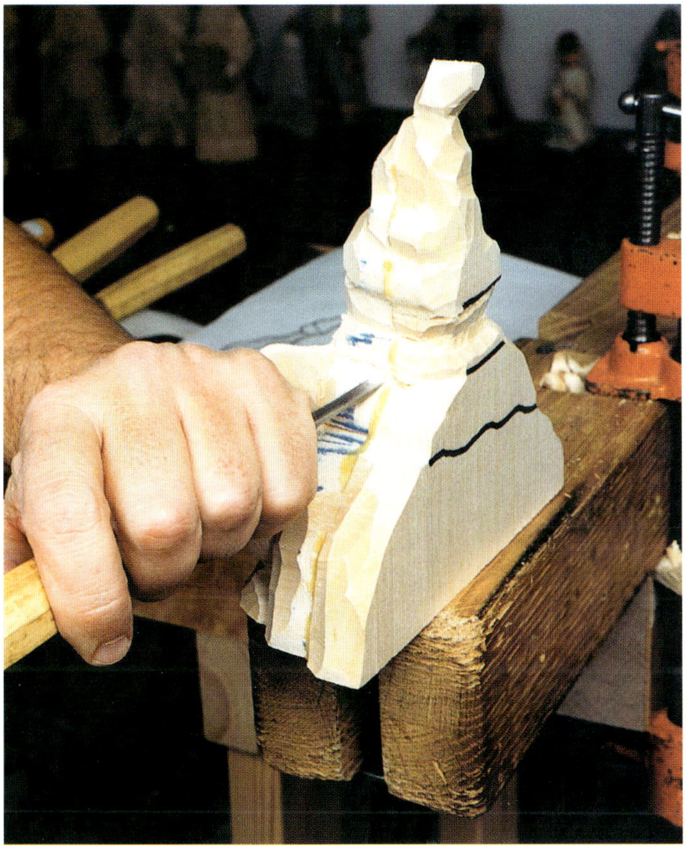

Move to the other side and cut up under the scarf.

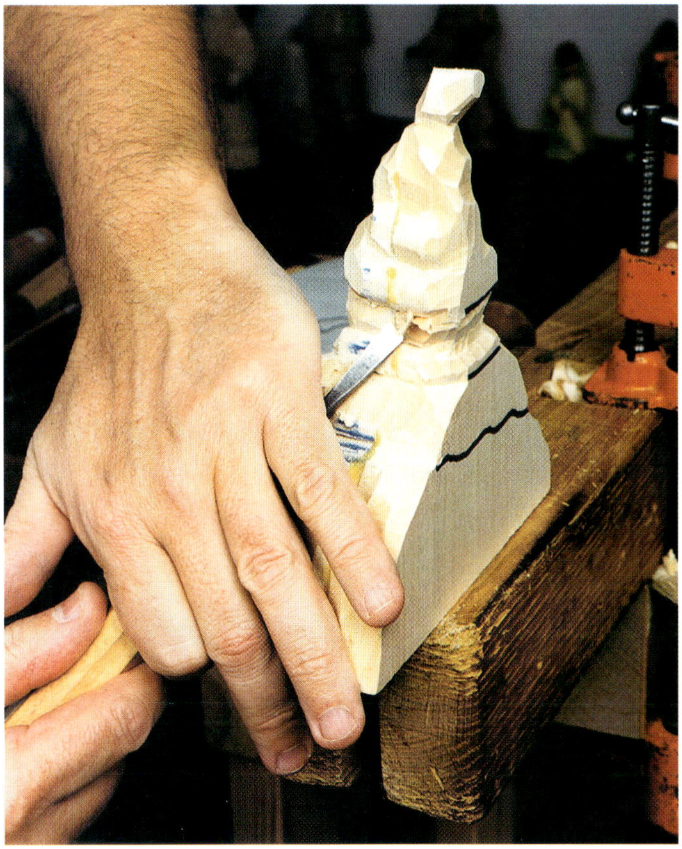

Shape the side of the head.

Shape the side and start around the back.

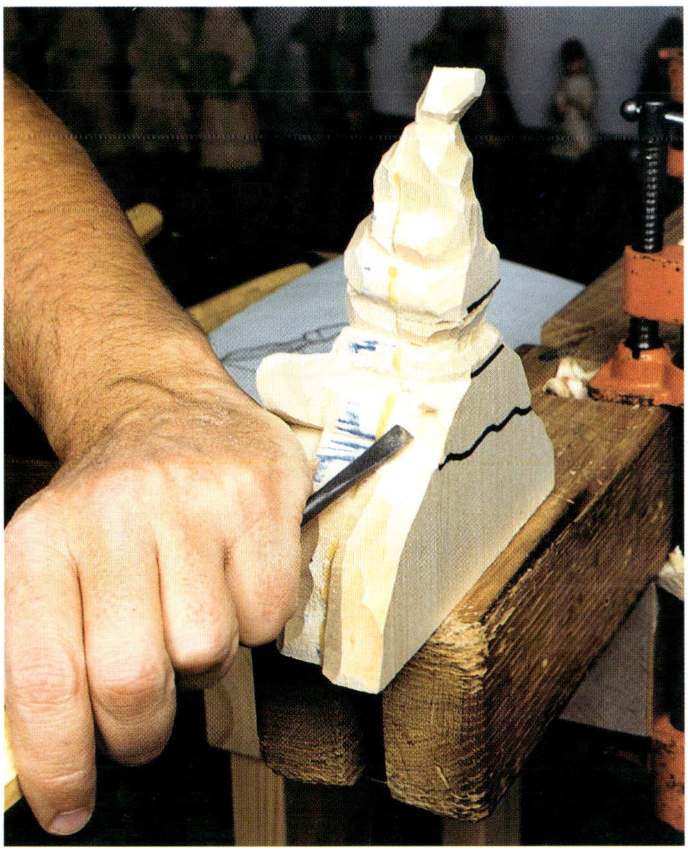

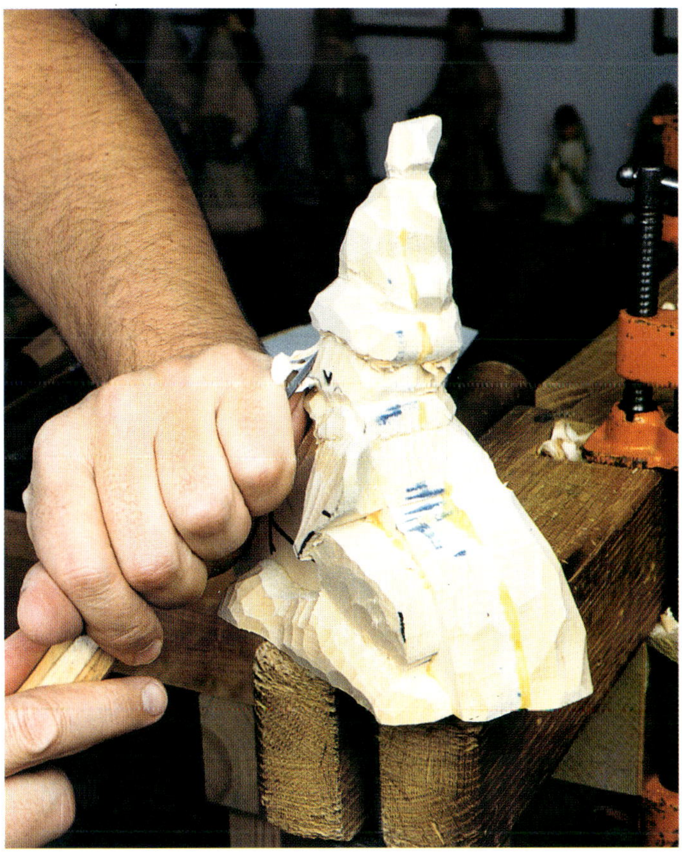

Work the other side of the head too.

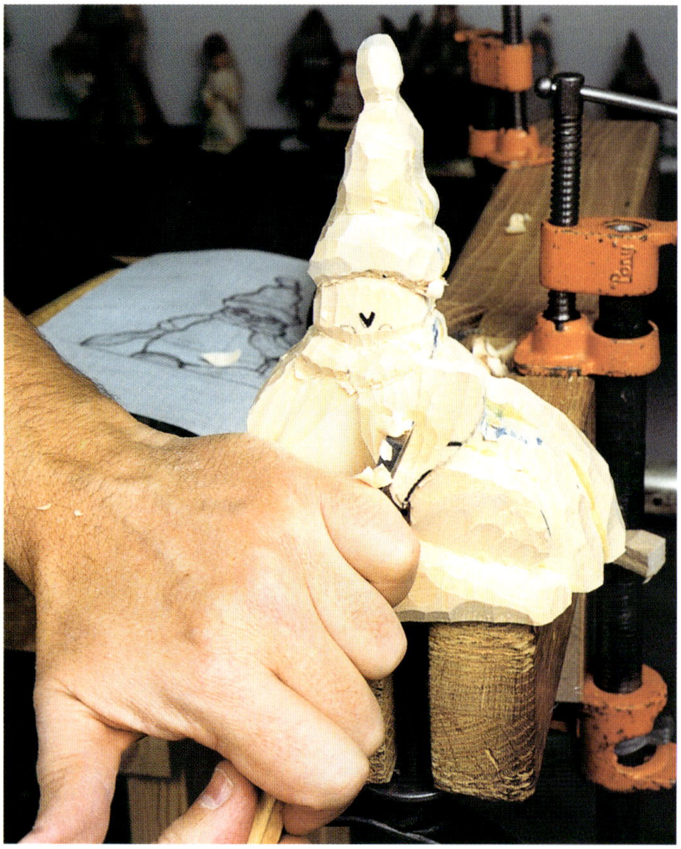

Shape the scarf the way you want to see it.

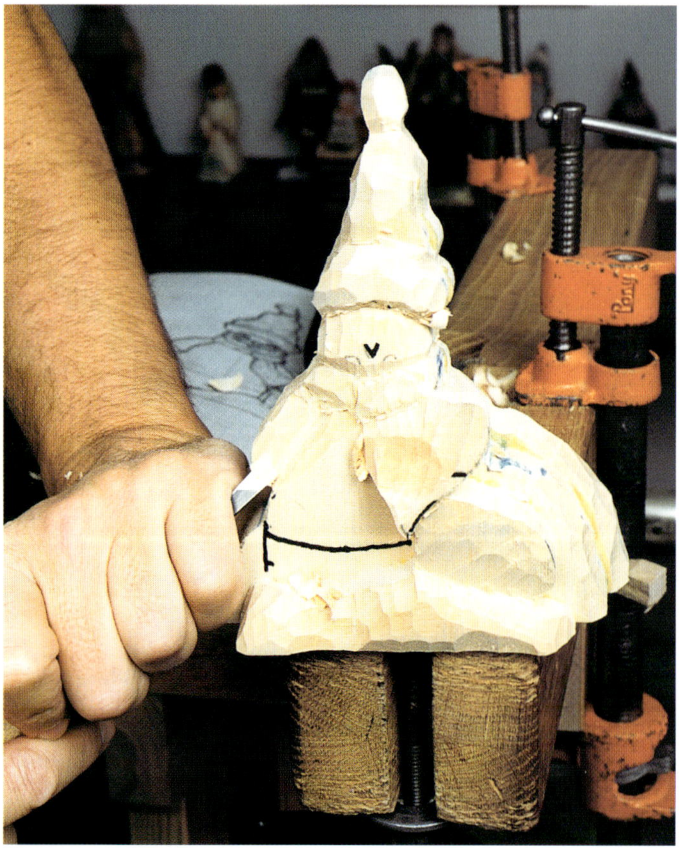

Shape the area in front of the arm.

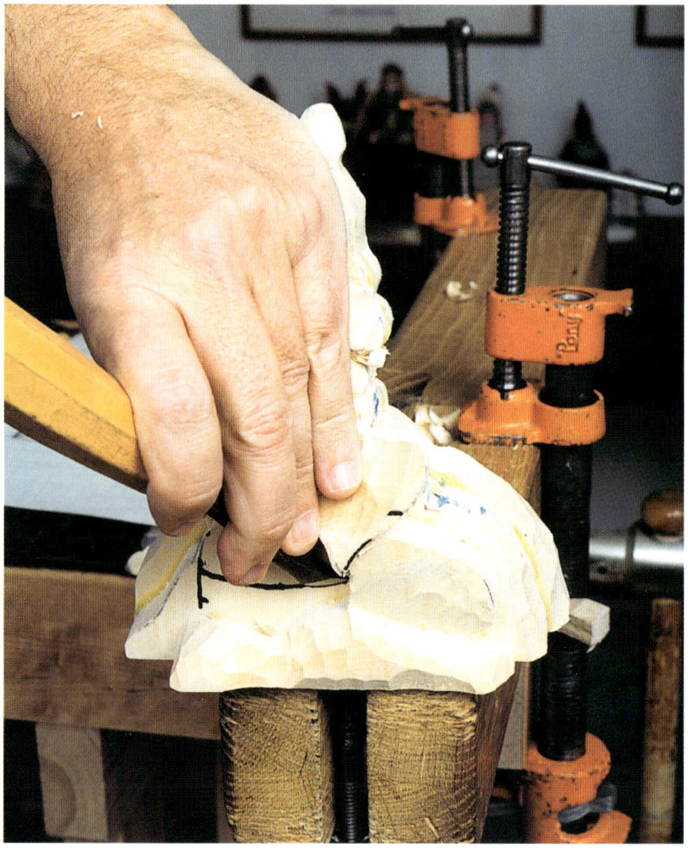

Make a stop cut along the bottom of the Snowman's coat.

Cut up to the bottom of the stop cut.

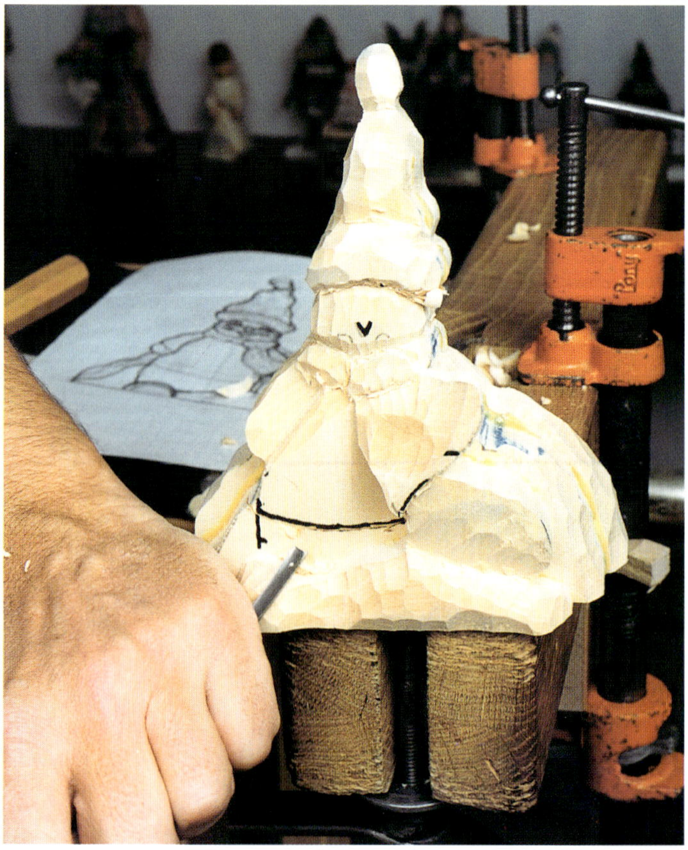

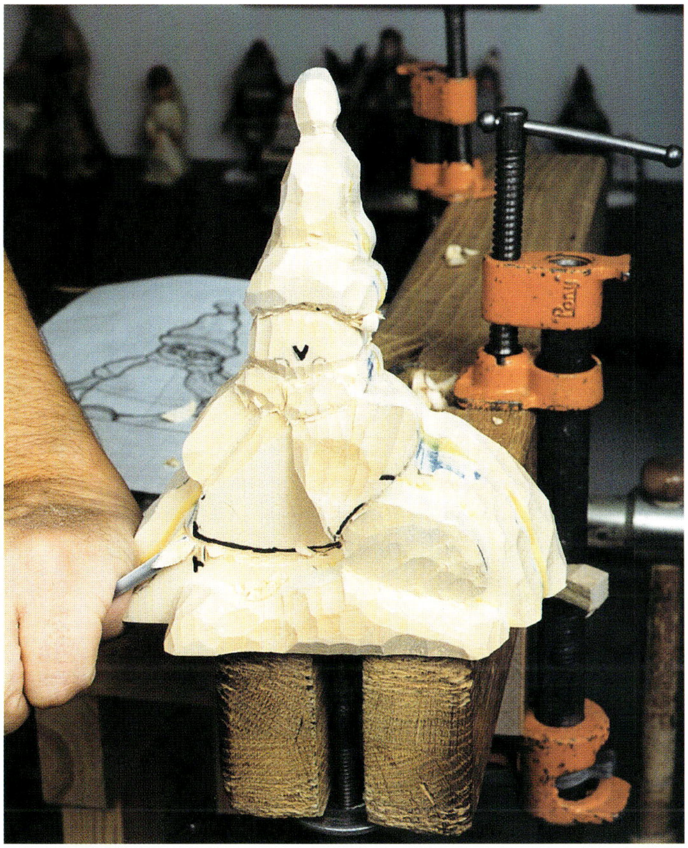

Cut up to the stop cut at the side.

Shape the area in front of the arm once more.

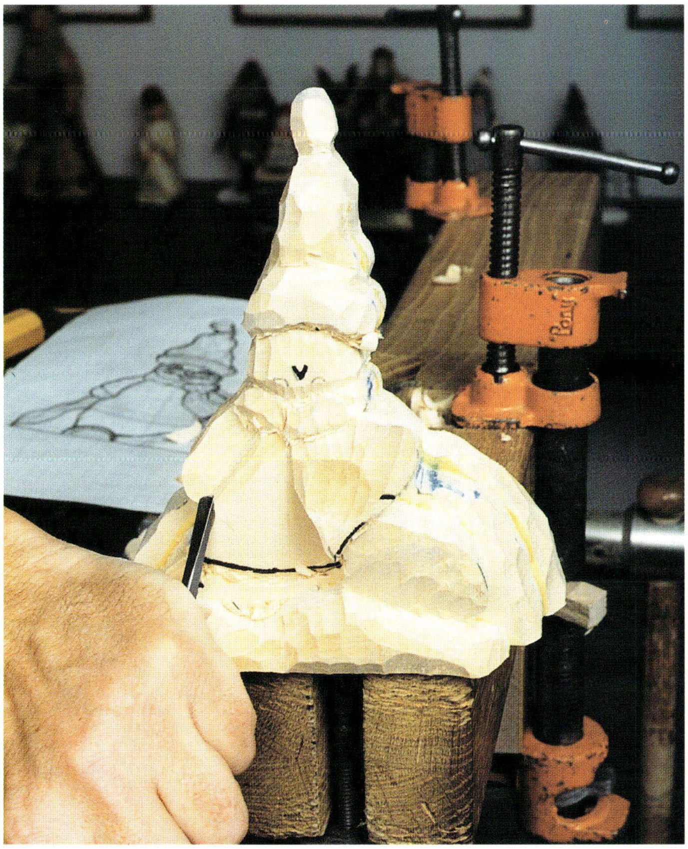

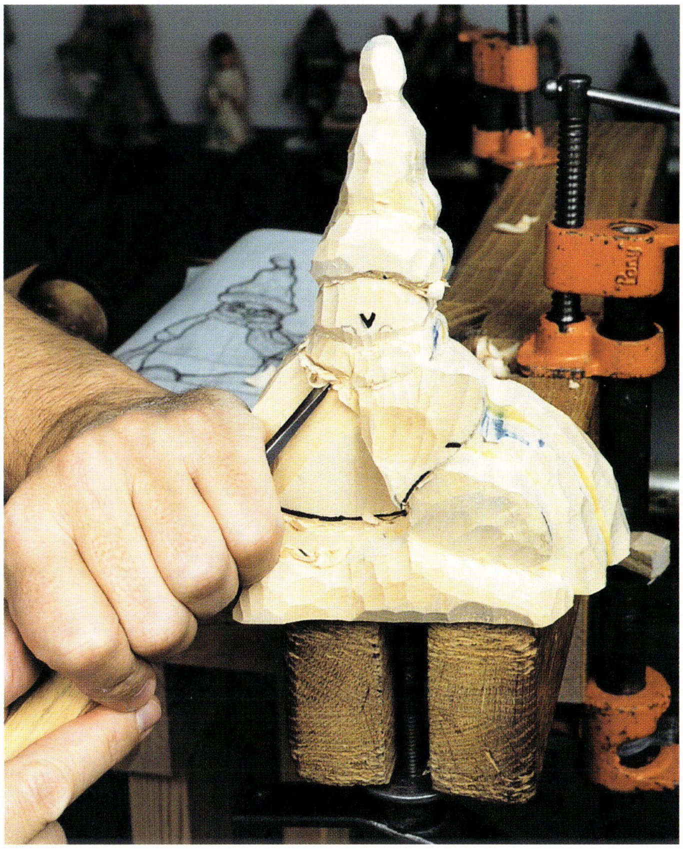

Continue shaping the upper body.

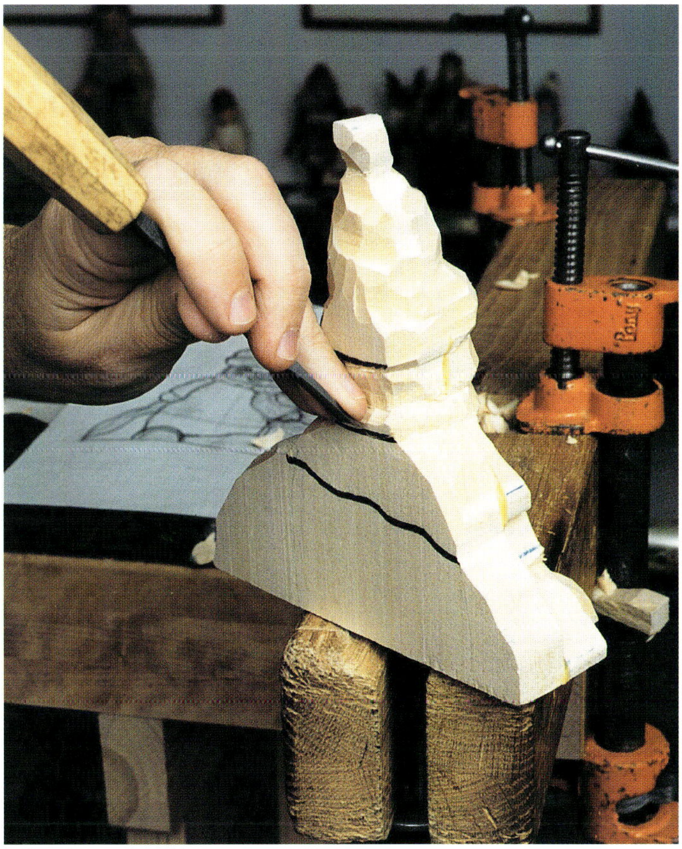

Re-cut the stop cut along the bottom of the scarf in the back.

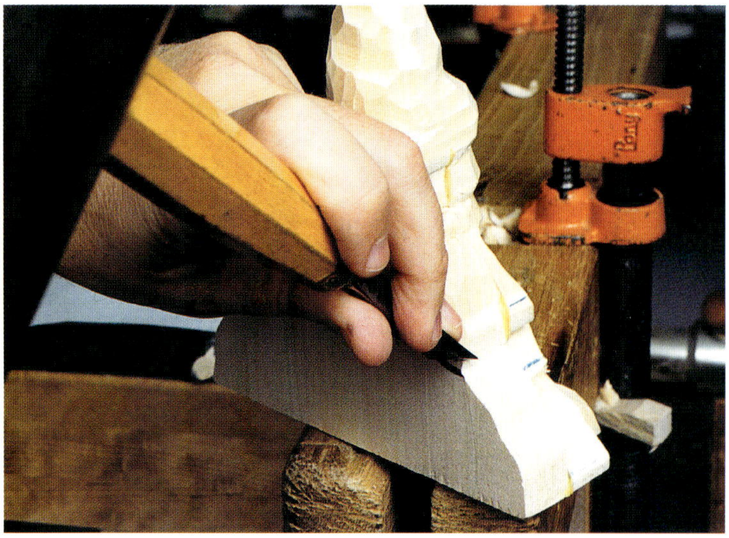

Make a stop cut at the top of the snowbank on the side of the piece.

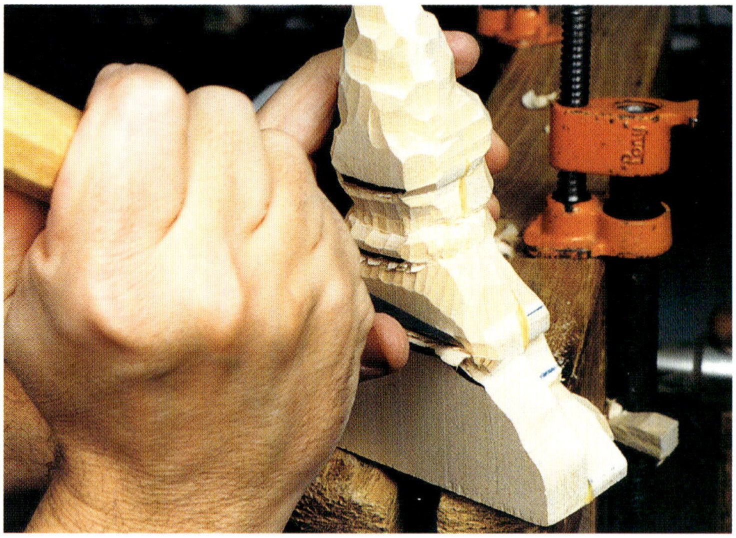

Make a stop cut at the top of the snowbank on the side of the piece.

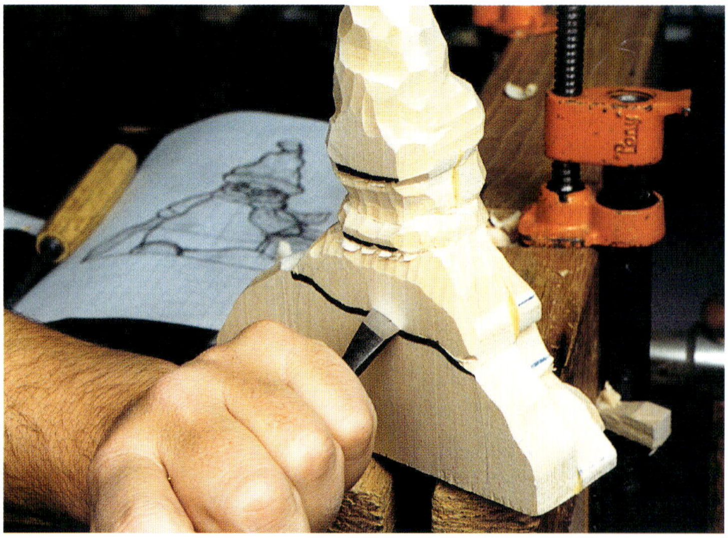

Put a stop cut along the top of the snowbank on the back.

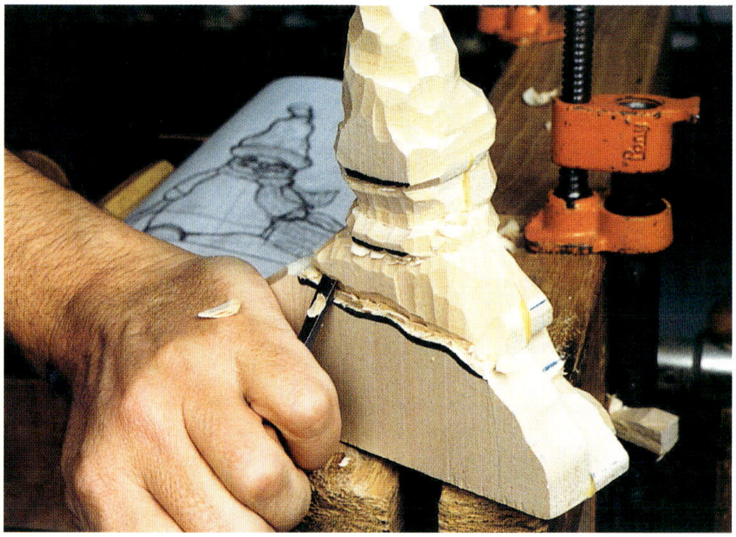

Cut up to the snowbank stop cut.

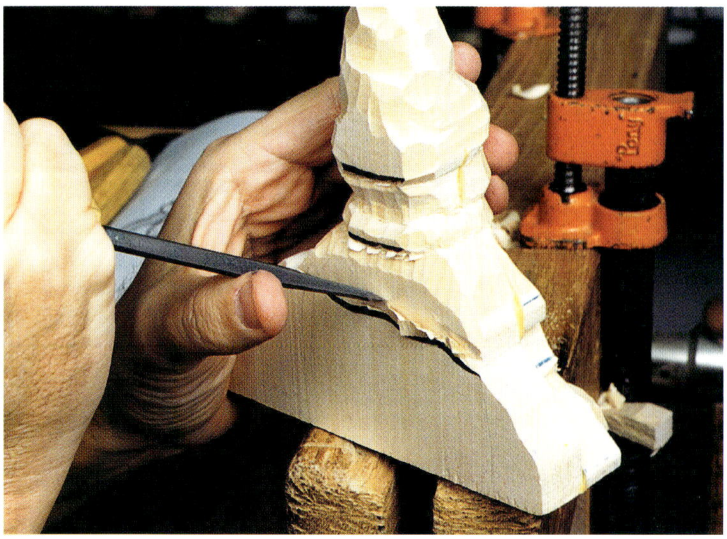

Cut down to the top of the snowbank.

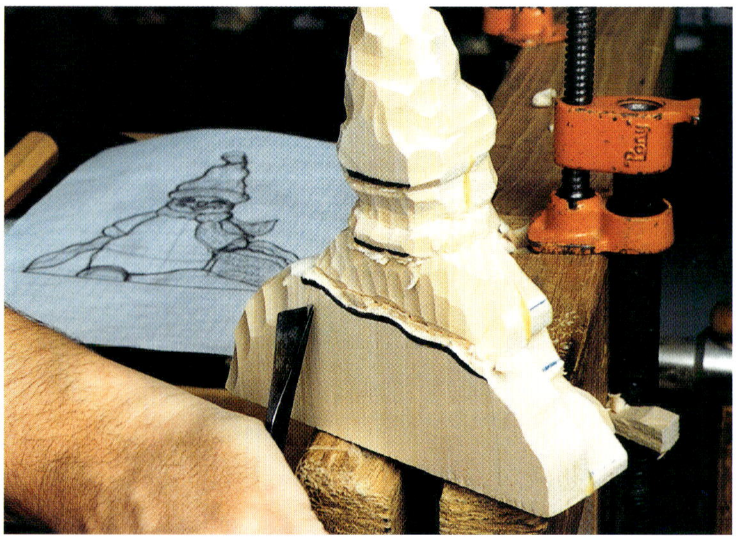

Use a large shallow "U" gouge to shape the back of the snowbank.

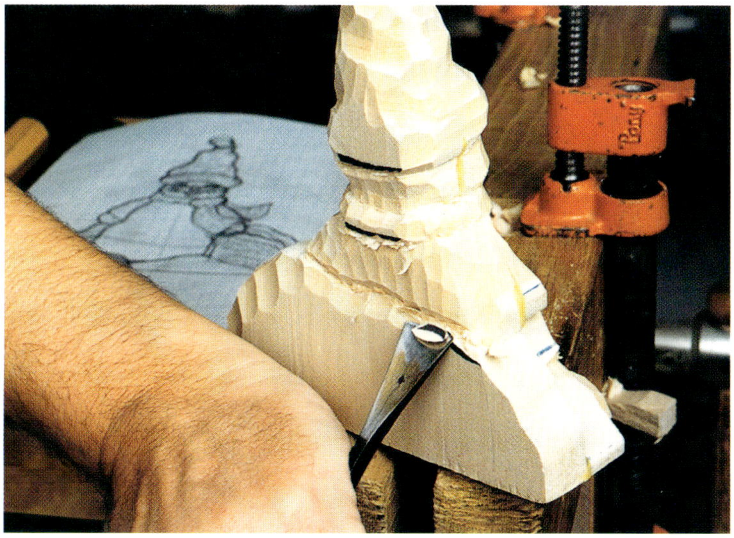

Shape the top of the snowbank.

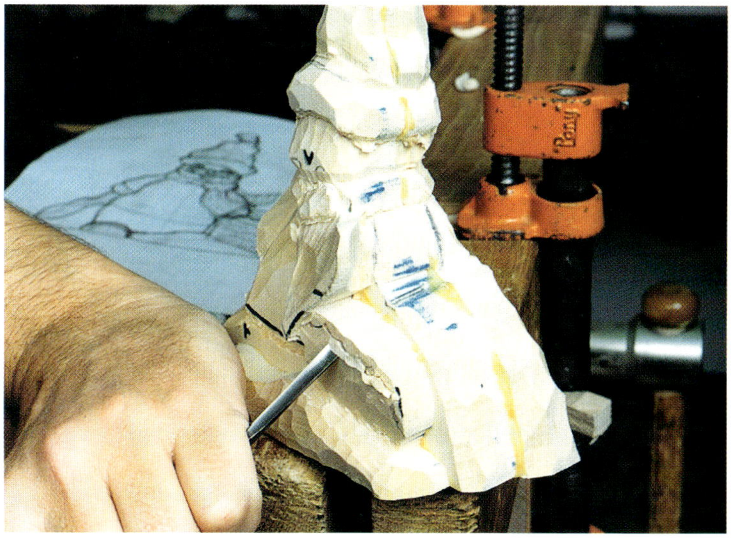

Cut up the face of the sign to the stop cut.

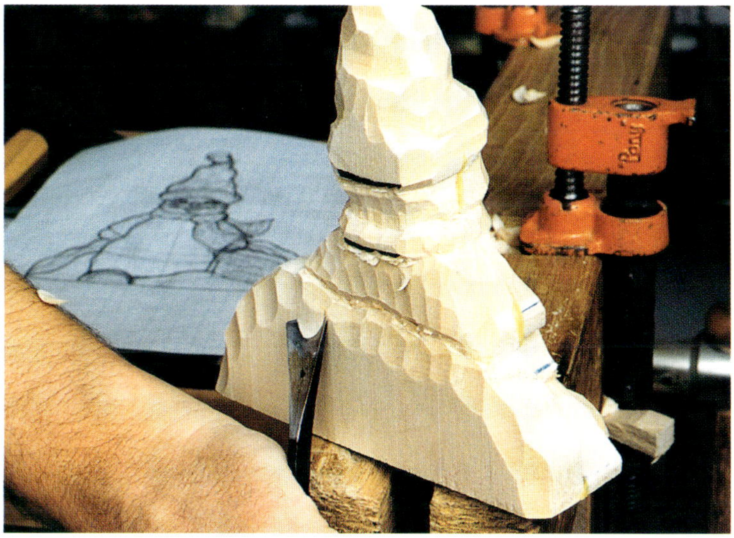

Shape the snowbank.

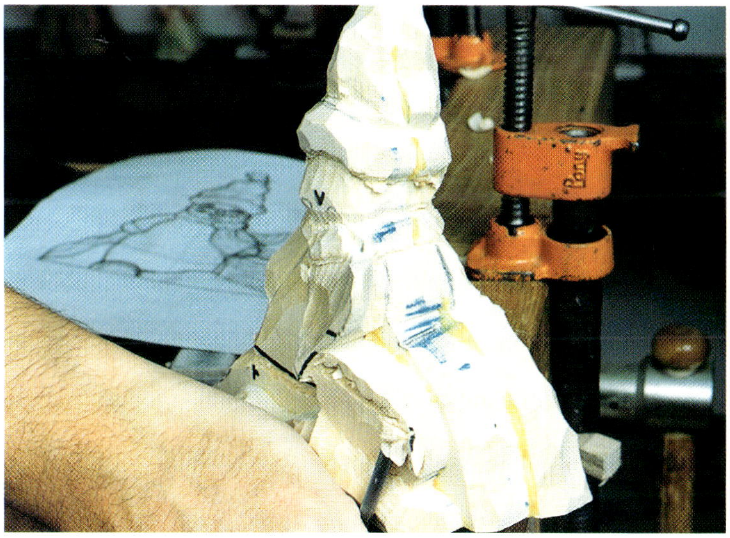

Cut the face of the sign up to the bottom of the stop cut.

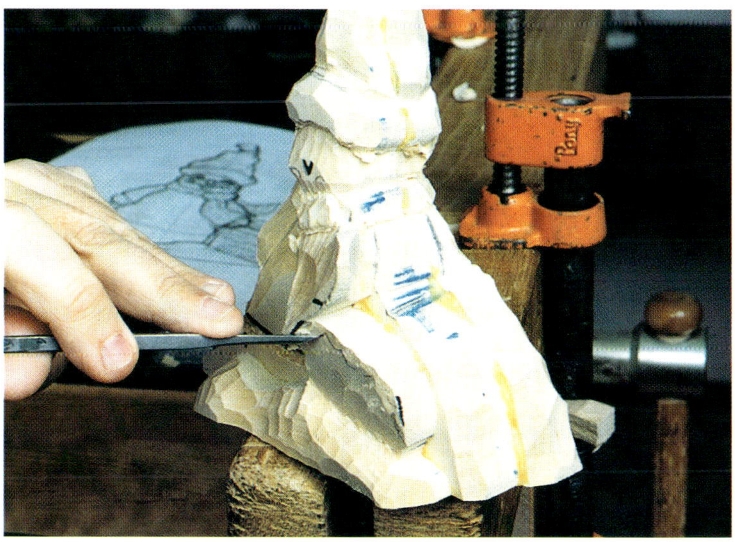

Put a stop cut on the front of the sign that defines the bottom of the "snow" sitting atop the sign.

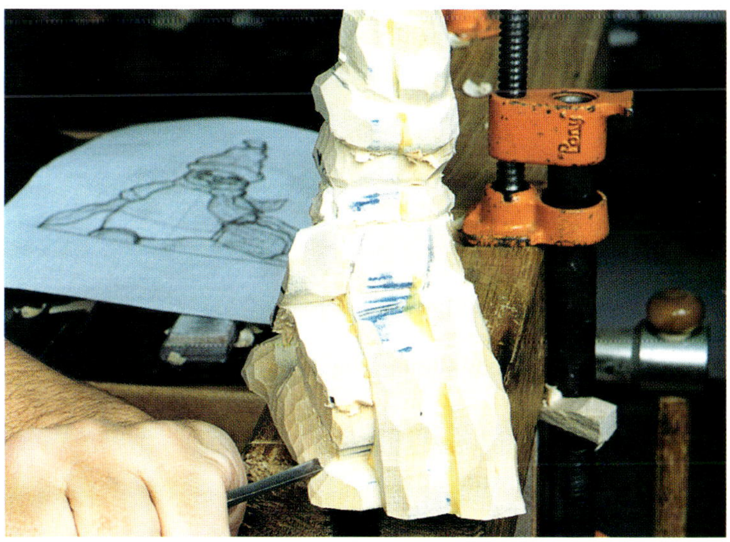

Shape the snowbank up under the sign.

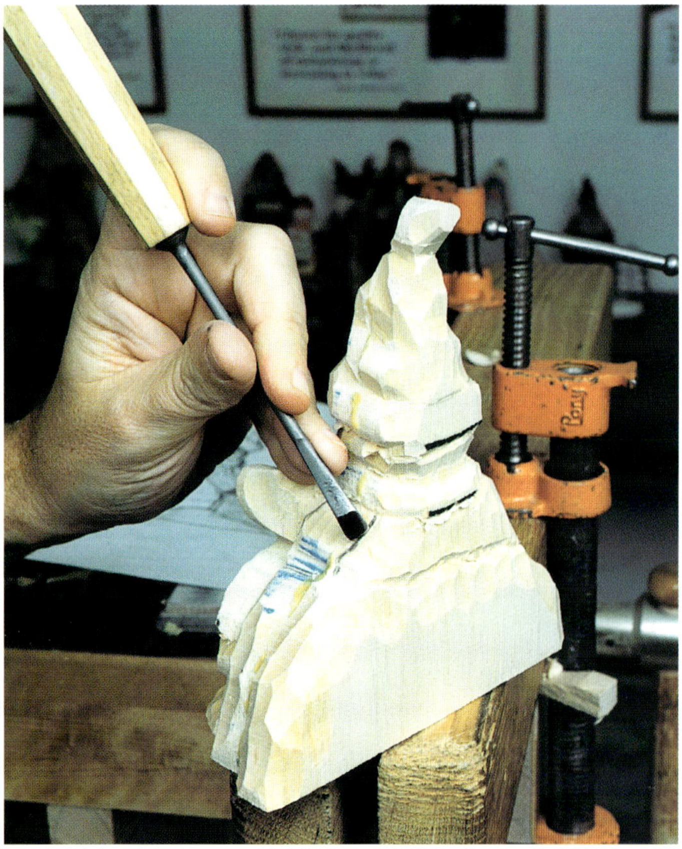

Define the back of the arm with a stop cut. The arm curves to the front and goes in back of the sign.

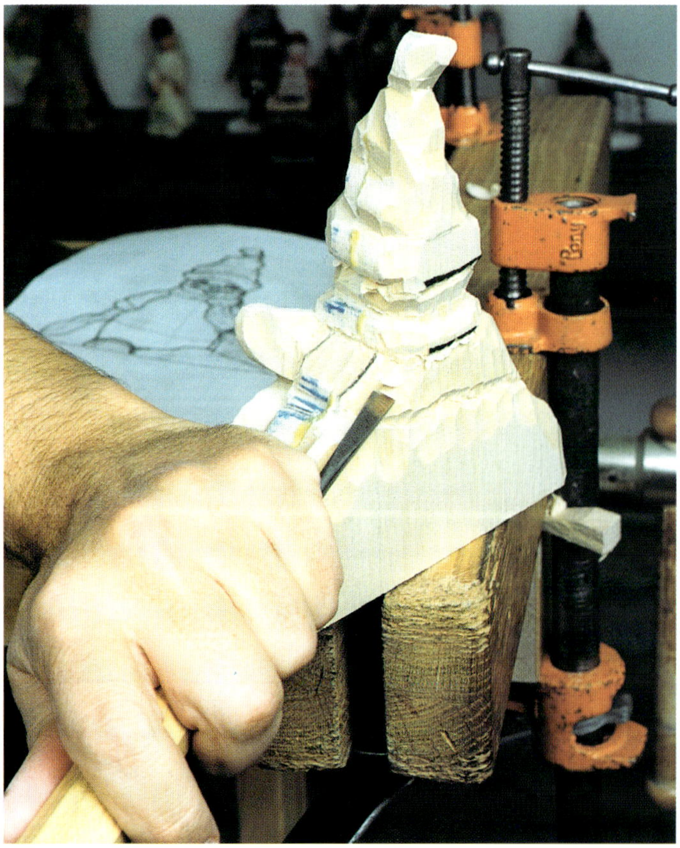

Cut away the wood behind the arm.

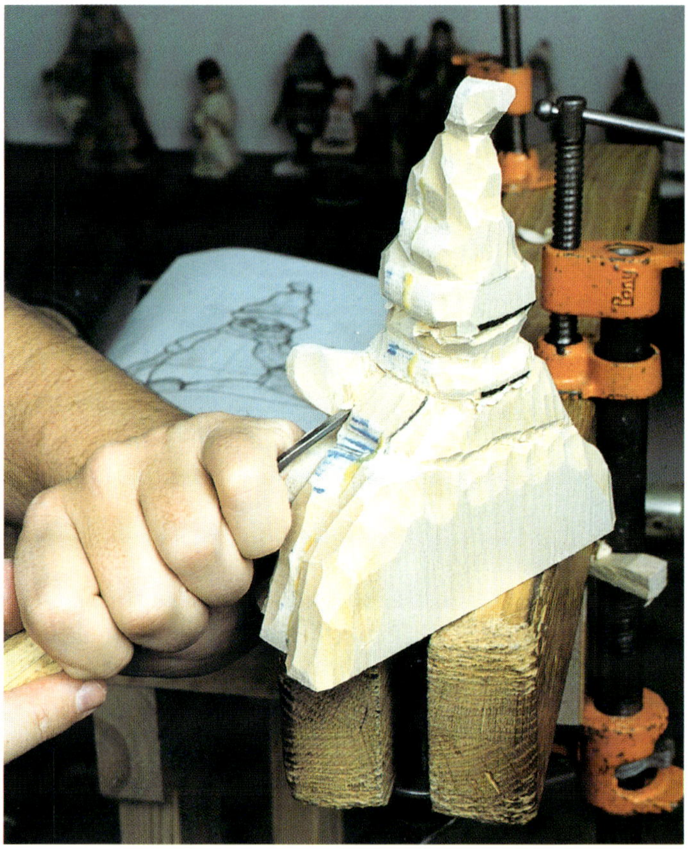

Clean up the back of the scarf.

Shape the back of the head.

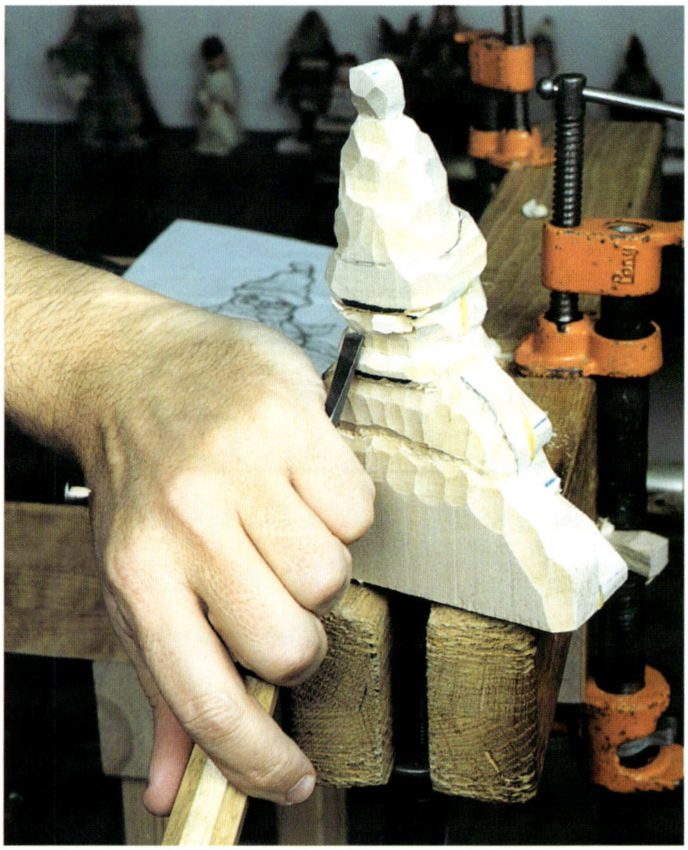

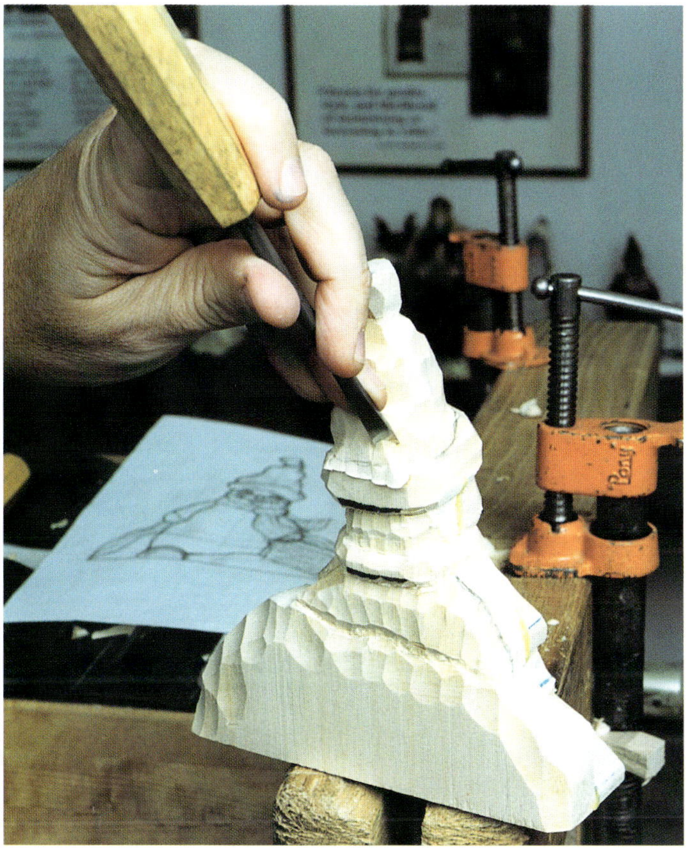

Shape the back of the hat.

Shape the back of the hat.

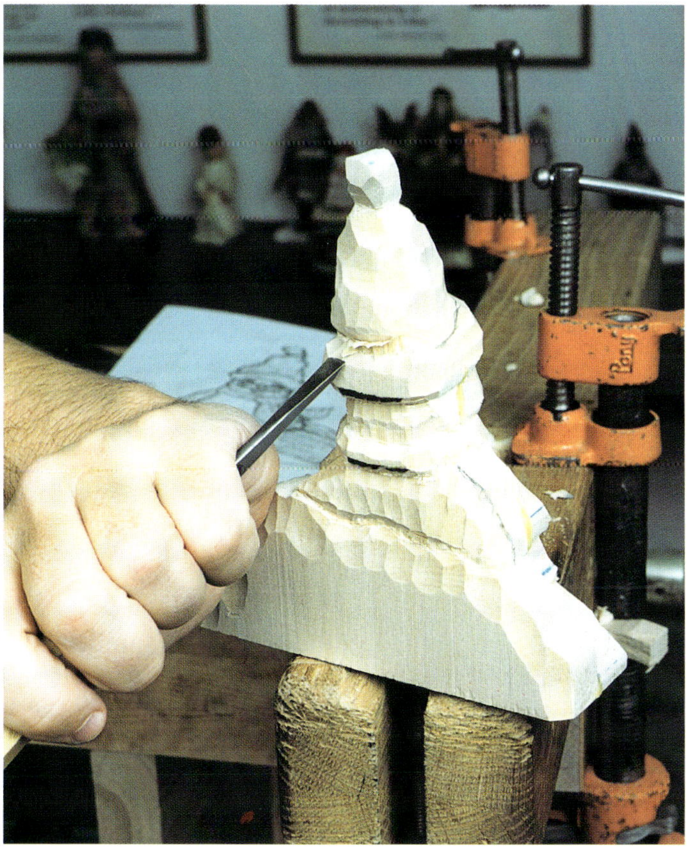

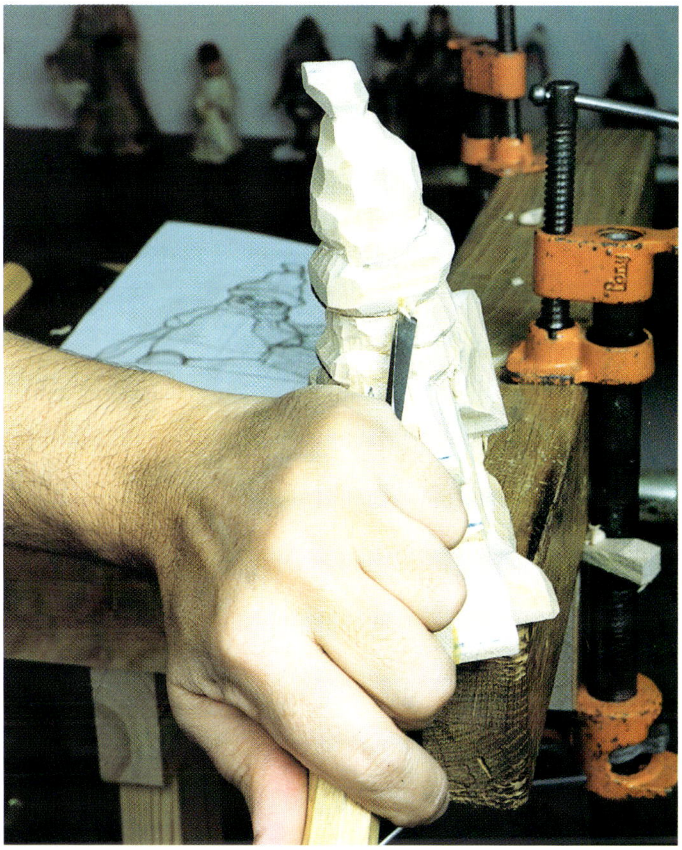

Shape the side of the head and hat.

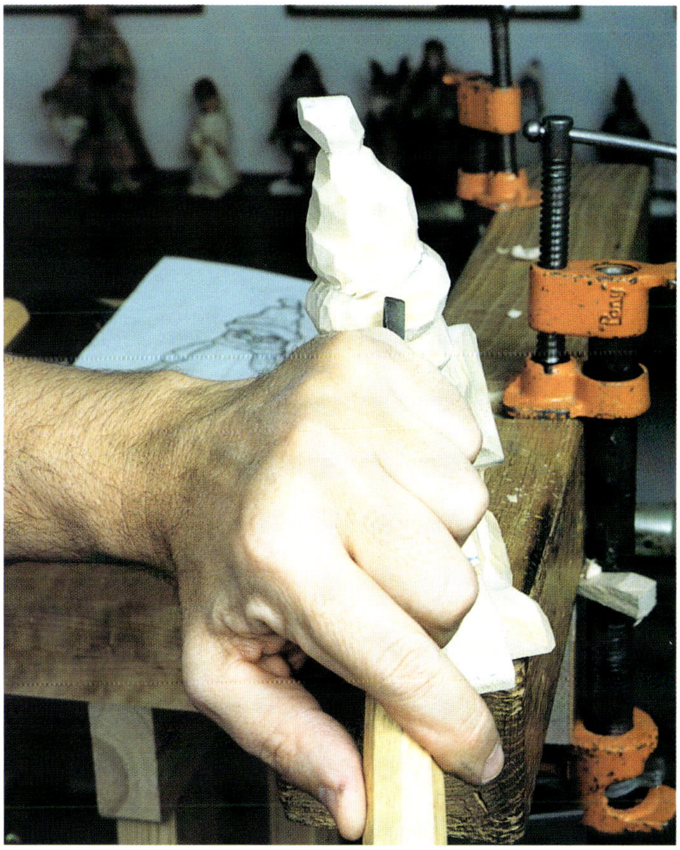

Shape the side of the hat.

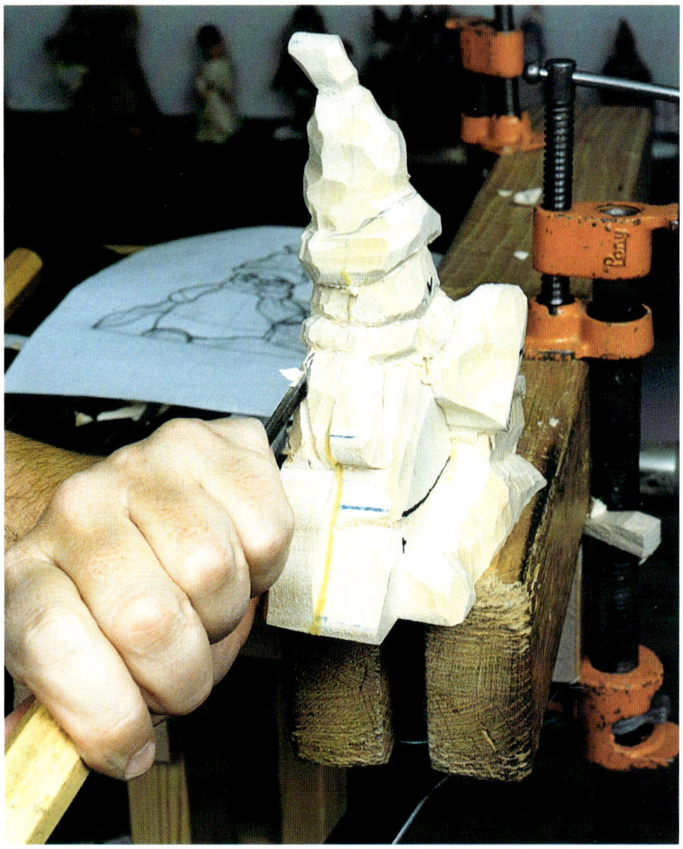

Shape the back of the right arm.

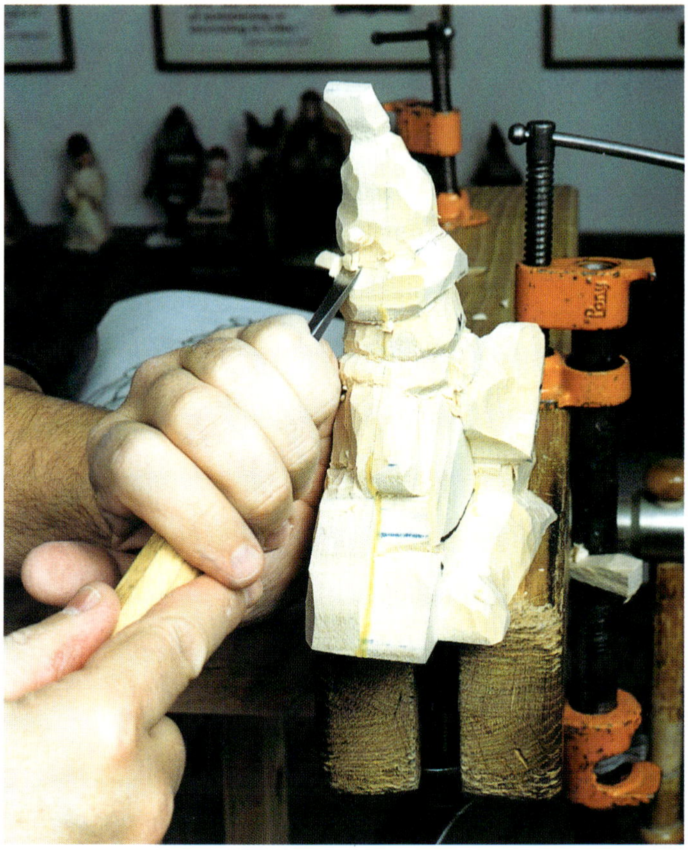

Shape the hat more. Pick your own shape.

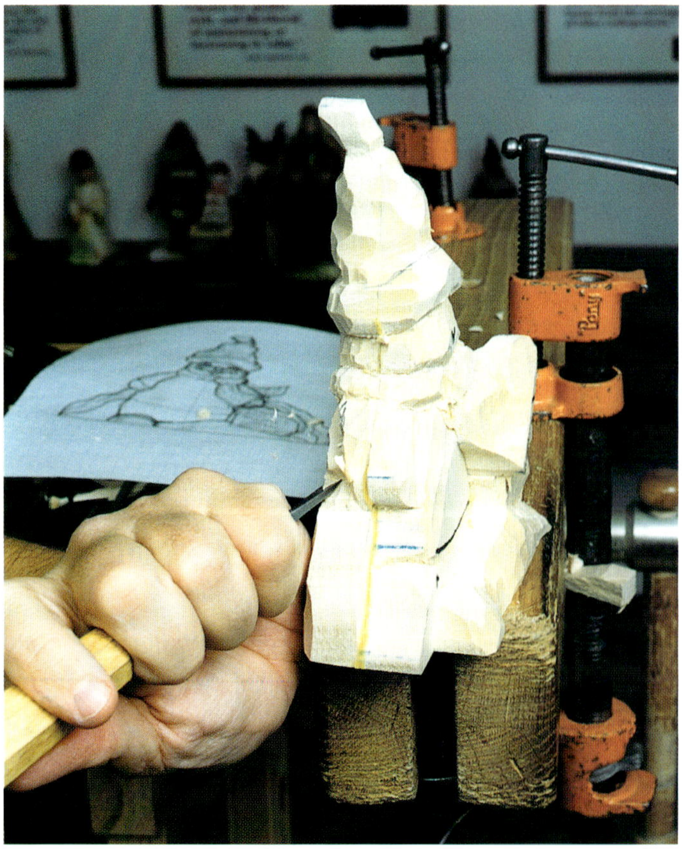

Deepen the back of the cut between the arm and the body.

Shape the back of the Snowman just above the snowbank.

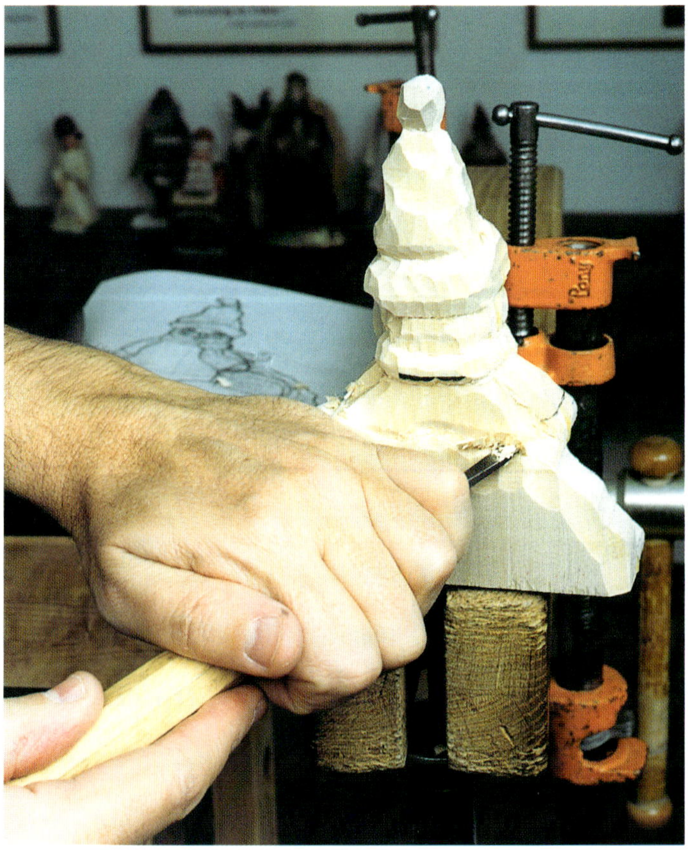

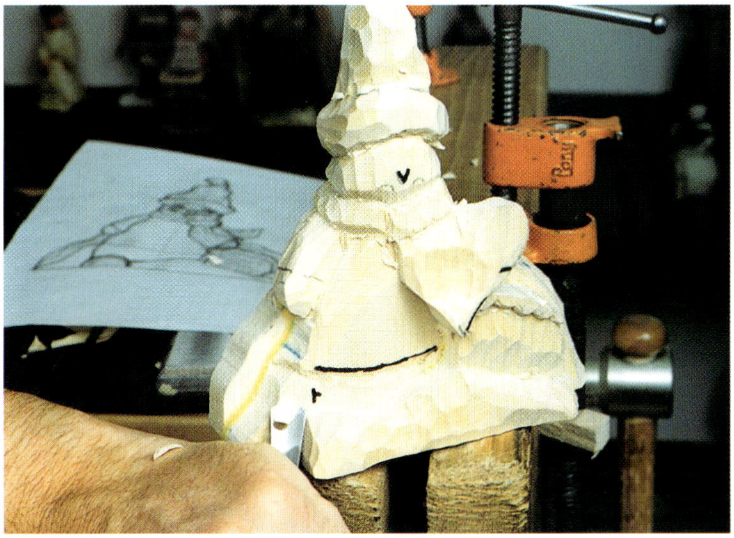

Shape the side of the lower snowbank.

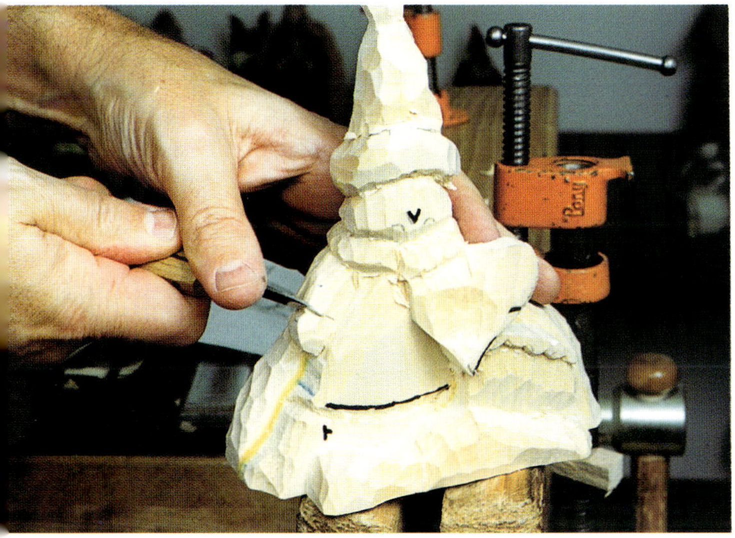

Make a cut separating the hand from the sleeve.

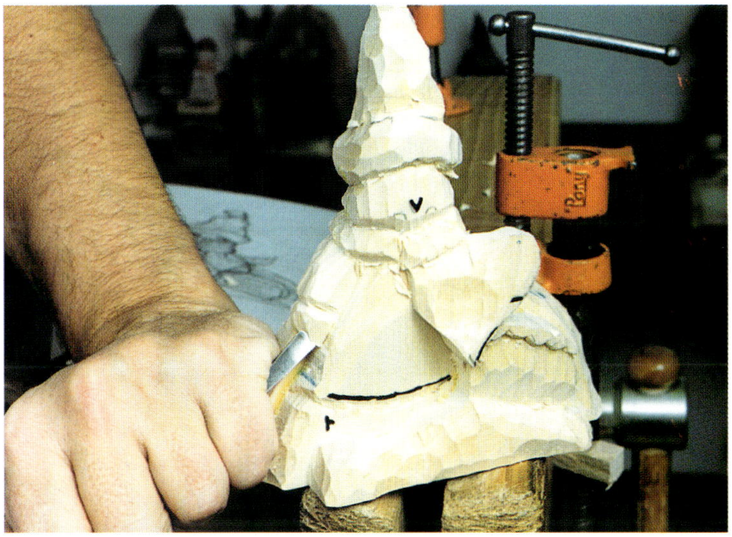

Define the hand and make it more round.

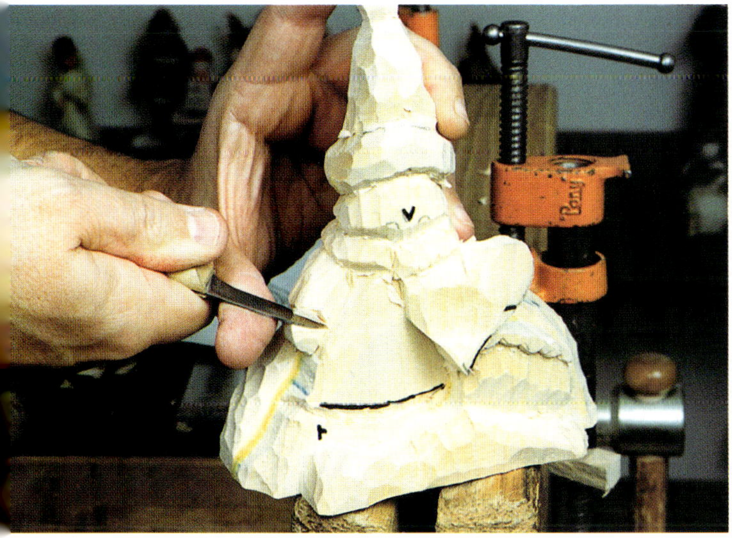

Deepen the cuts so the hand is separated from the sleeve.

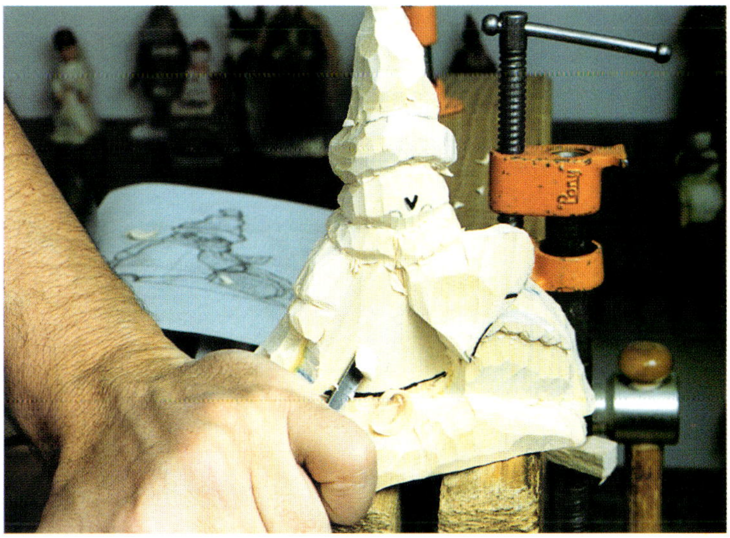

Shape the tummy of the Snowman.

From here you clean up the carving and do any final shaping that needs to be done. I use a Foredom rotary tool to clean up the carving and do some detailed shaping. After that, the piece is hand sanded to get it smooth enough for painting.

Chapter 5

Painting the Snowmen

Camille painted the snowmen using acrylic craft paints, straight from the bottle, to make it easy. She picked some pretty wild colors, but, in the world of Snowmen, anything goes. I guess there is no such thing as a conservative snowman. If you remember the discussion in my other books about the use of color, you will already know that color is the first thing that people tend to see. If you want to catch someone's eye, use color. Shape, size, design, and the other factors are important, but color comes first.

The carved and sanded pieces are base-coated with a product called "White Lightning," which seals the wood and provides a base for painting. After the base coat is dry, Camille, or I, do one final hand sanding with super fine sandpaper to remove the last "fuzzies." The White Lightning helps raise these last "fuzzies" and makes it easier to find and eliminate them.

The acrylic paints dry relatively quickly so quite a bit of painting can be accomplished in one session. The paints are thick and cover well. You can thin them down for other effects but on these snowmen we wanted intense colors – nothing faded. I guess a real snowman would melt long before he ever faded.

Camille likes to paint a large area quickly and then come back to do the "cut-in" between areas. The smaller details are added to the piece in the final coat. Look at the pictures, follow the instructions, and you should recognize some nice Snowmen at the end of the process.

You are encouraged to experiment with your painting.

This is the front view of the Igloo Snowman after he has been base-coated with White Lightning. It is a little hard to see any of the carved details but they are there.

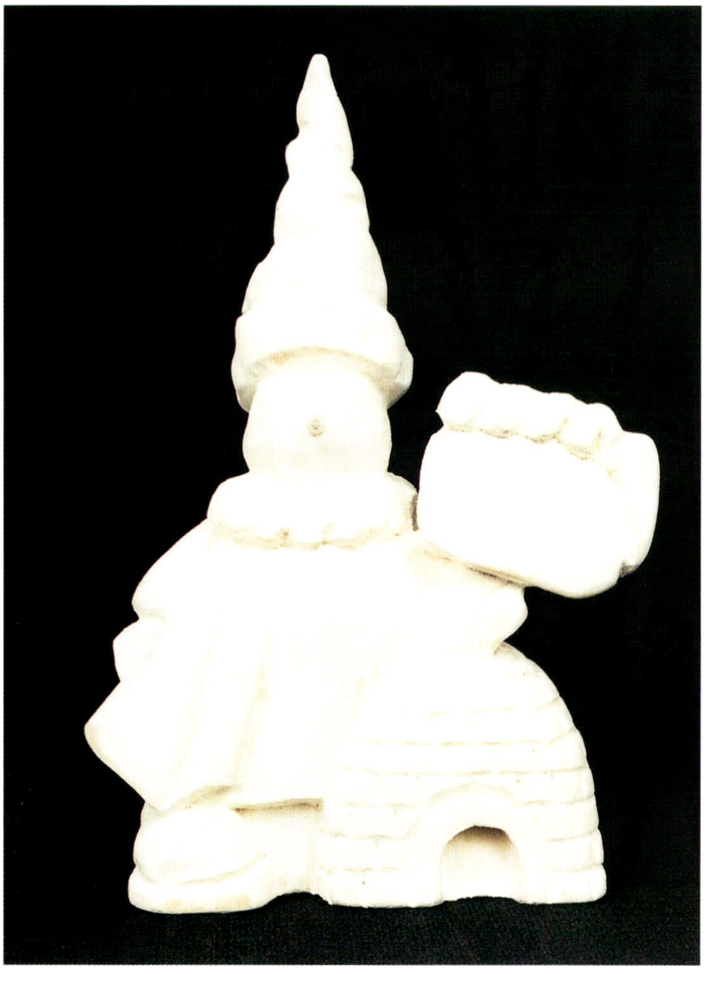

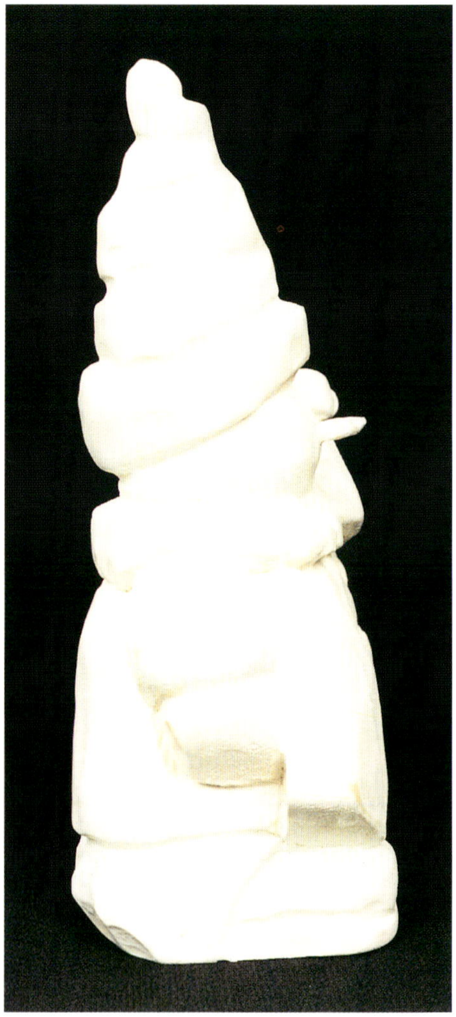

This is the right side view of the Igloo Snowman.

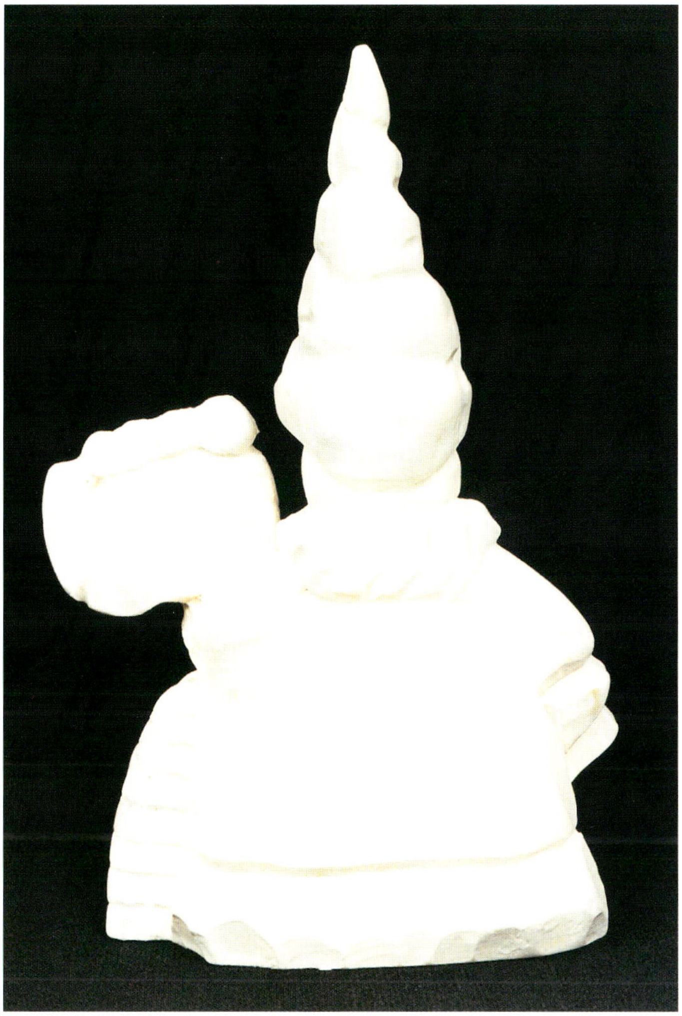

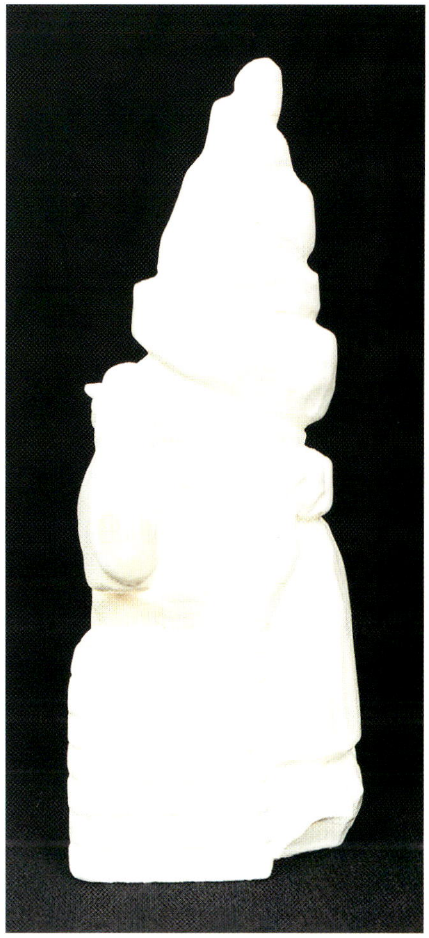

Right:
This is the left side view of the Igloo Snowman.

This is the back view of the Igloo Snowman.

This is the front view of the Snowbank Snowman.

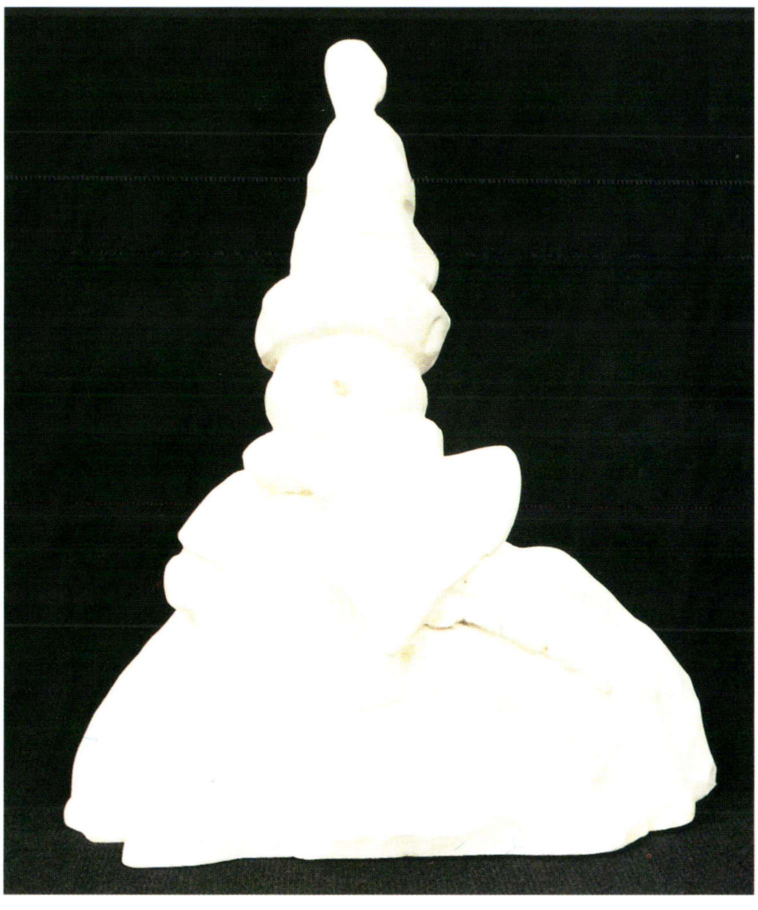

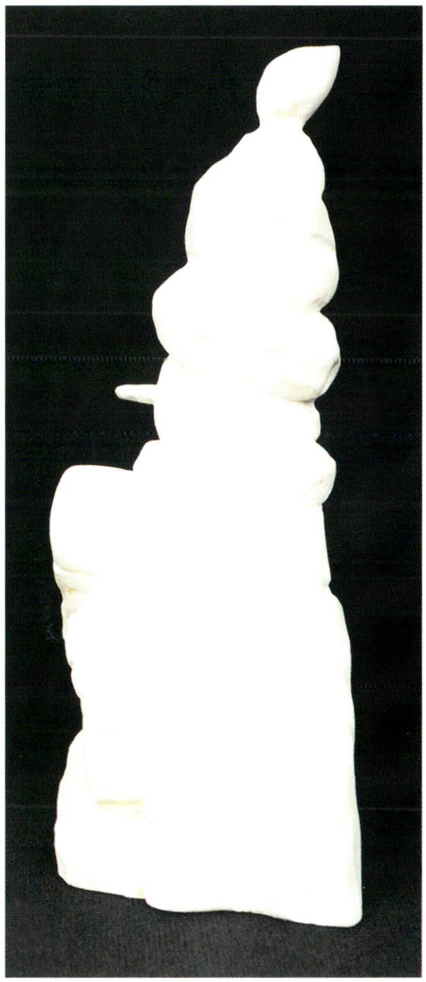

This is the left side view of the Snowbank Snowman.

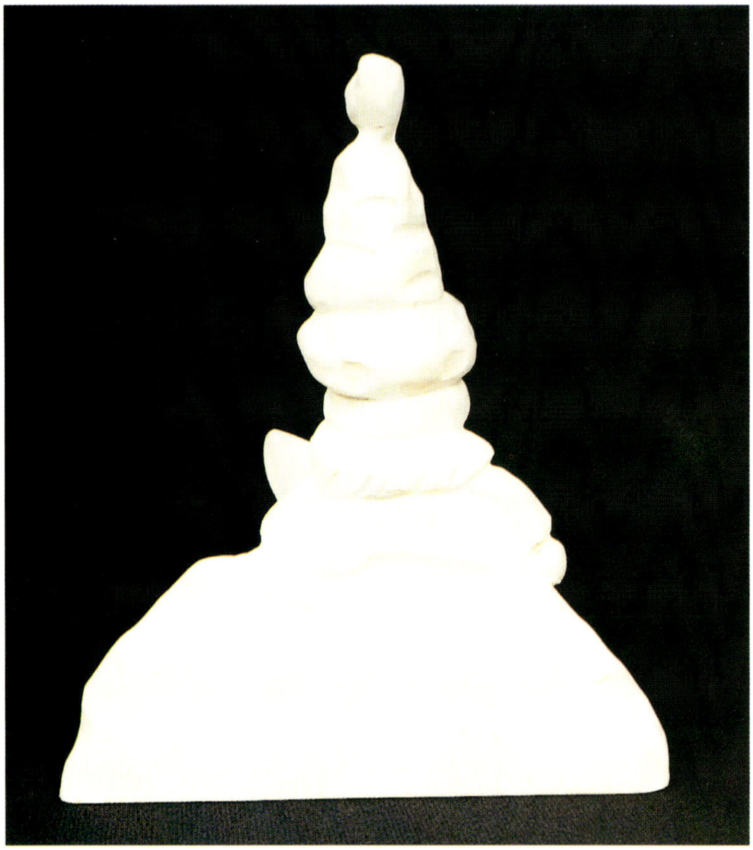

This is the back view of the Snowbank Snowman.

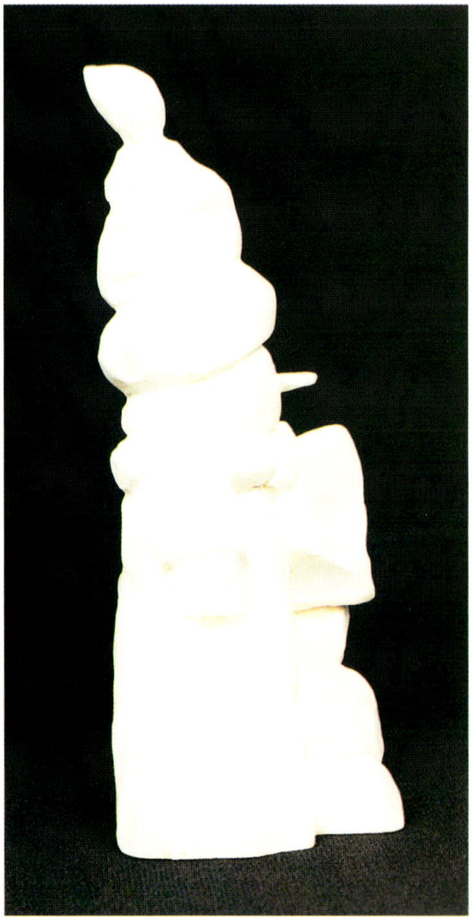

This is the right side view of the Snowbank Snowman.

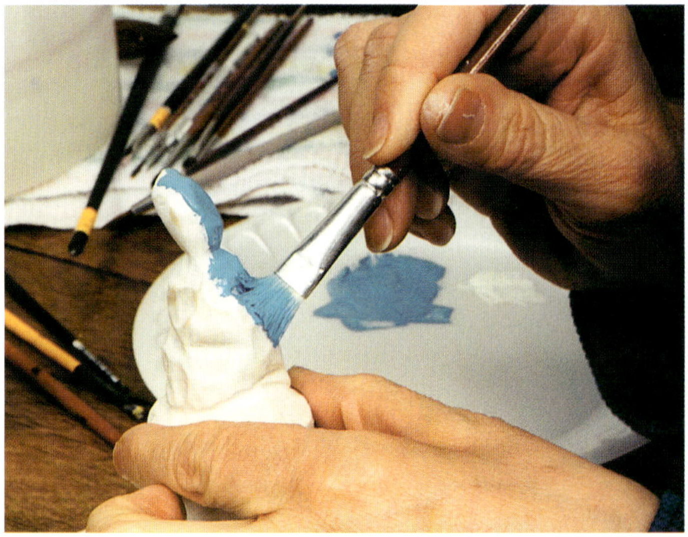

Camille started with the Snowbank Snowman. She decided he would have a bright blue coat. Blue is not a Christmas color but these snowmen are not about to be limited to the Christmas season.

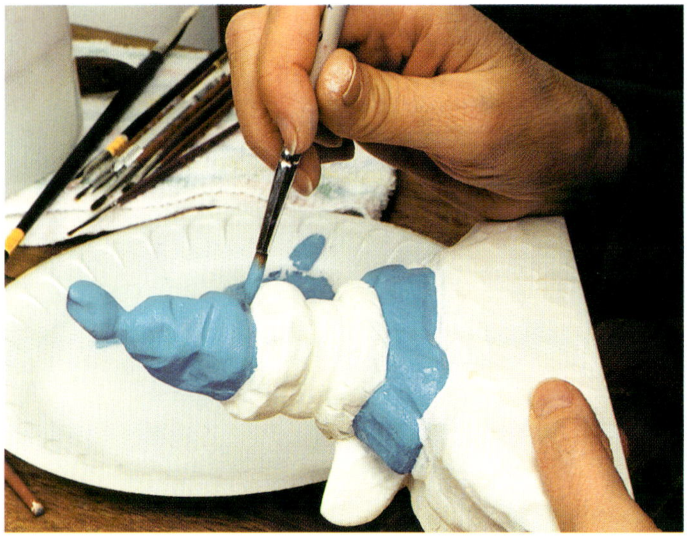

The hat will also be blue. Camille paints the larger areas quickly and then comes back to cut in more carefully.

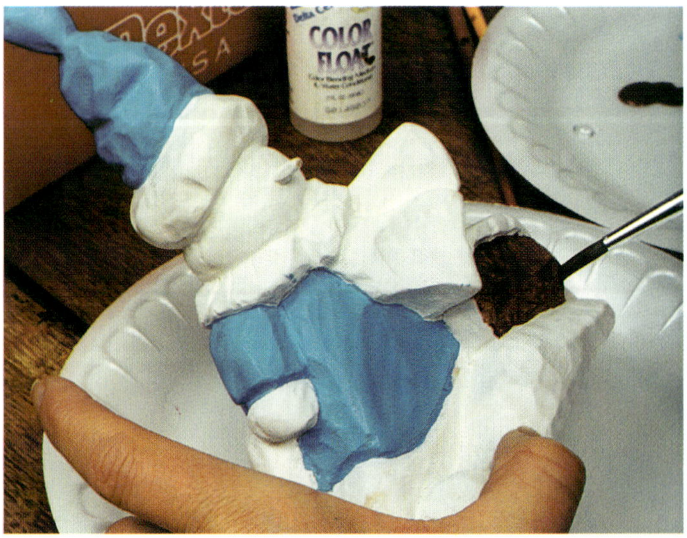

The front of the sign will be brown.

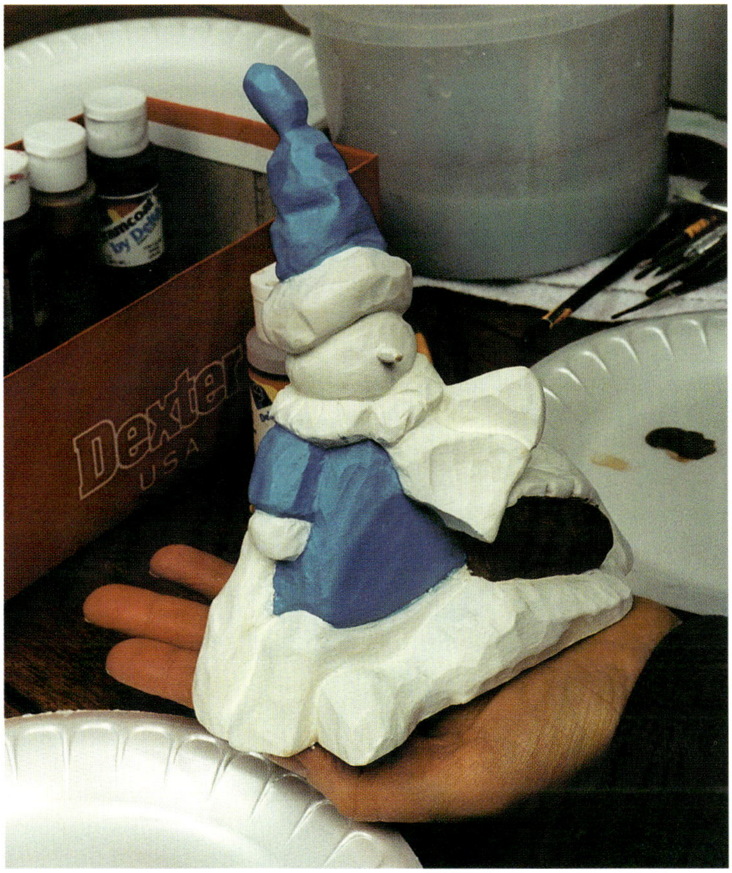

Front view of partially painted snowman.

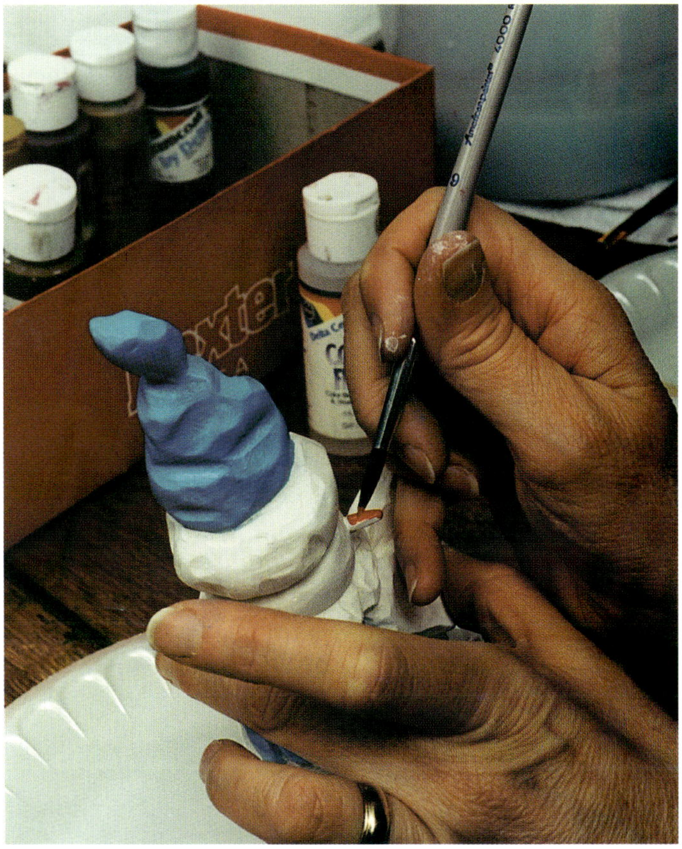

Paint the carrot nose an appropriate orange color.

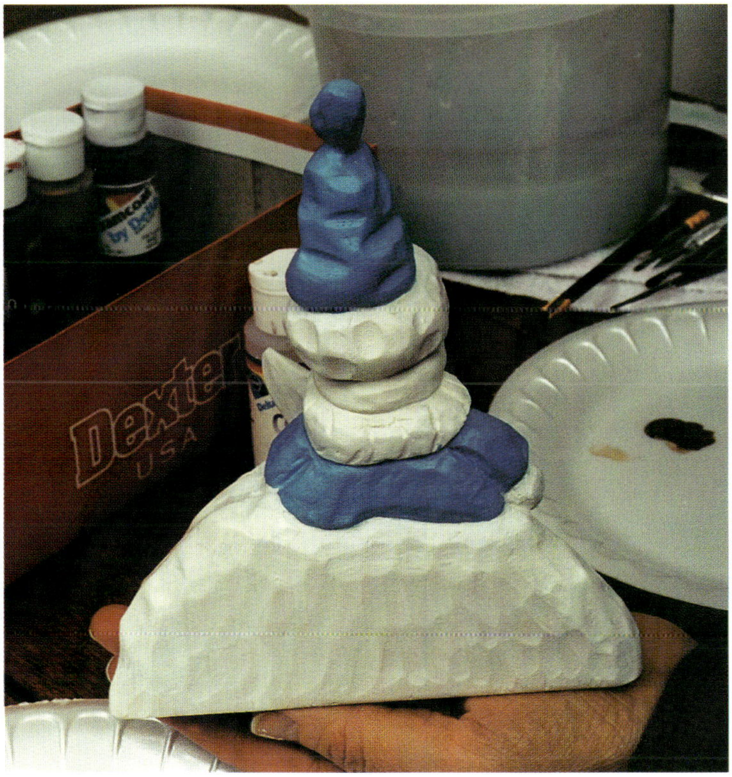

Rear view of snowman.

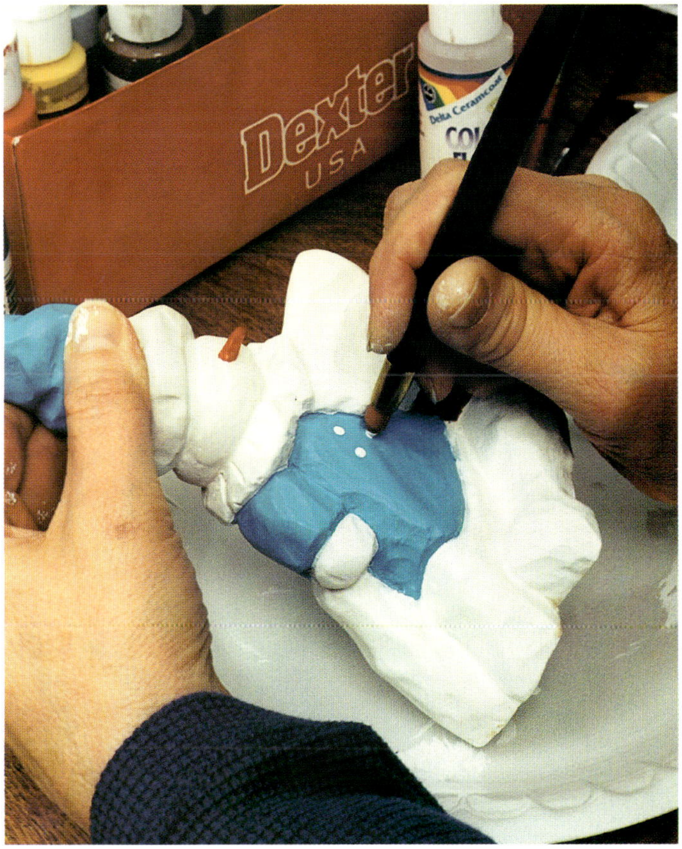

Using her rubber tipped stylus, Camille applies white dots over the coat.

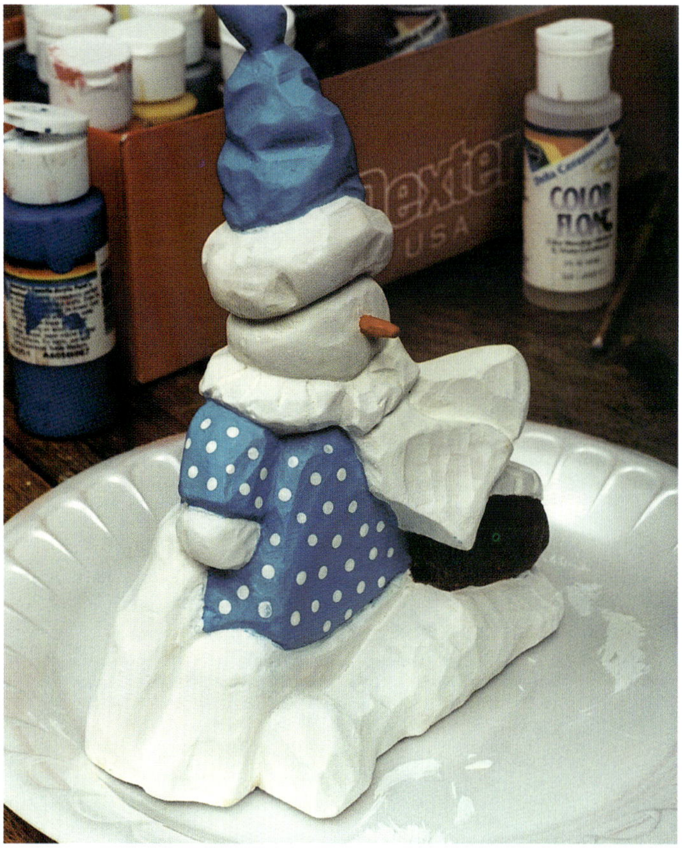

Here is the polka dot coat.

This is the back of the polka dot coat.

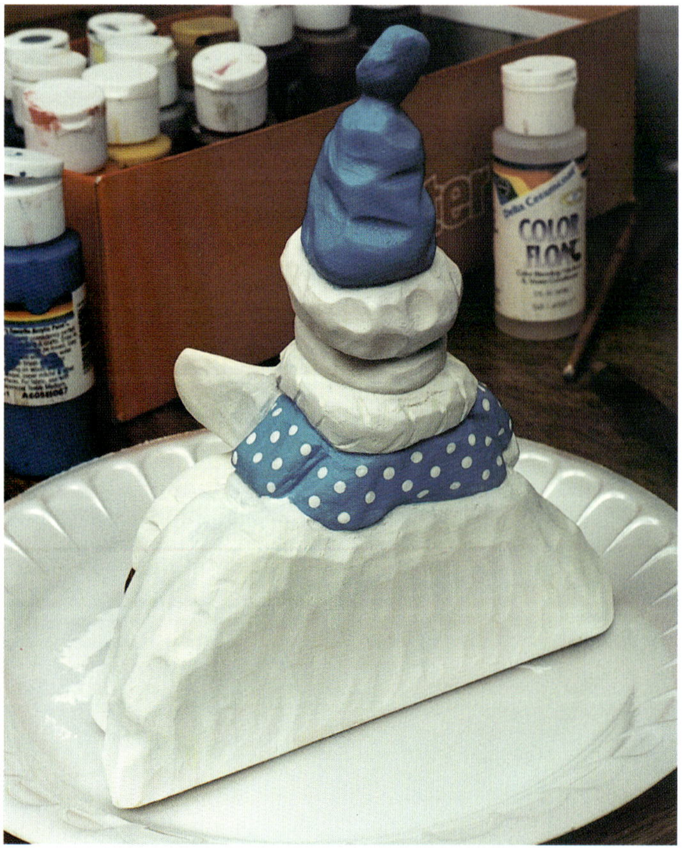

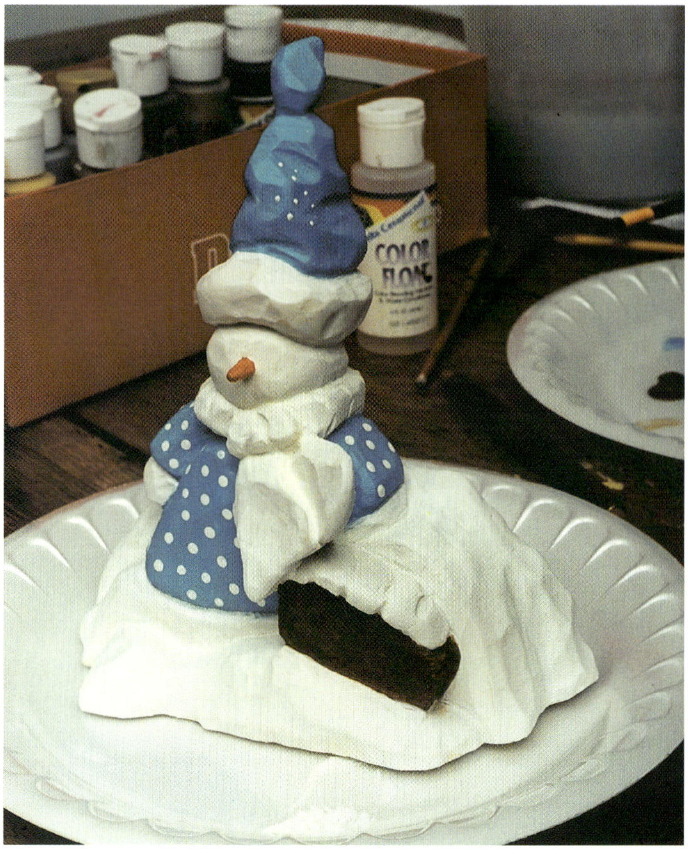

Camille has started to put smaller dots on the hat.

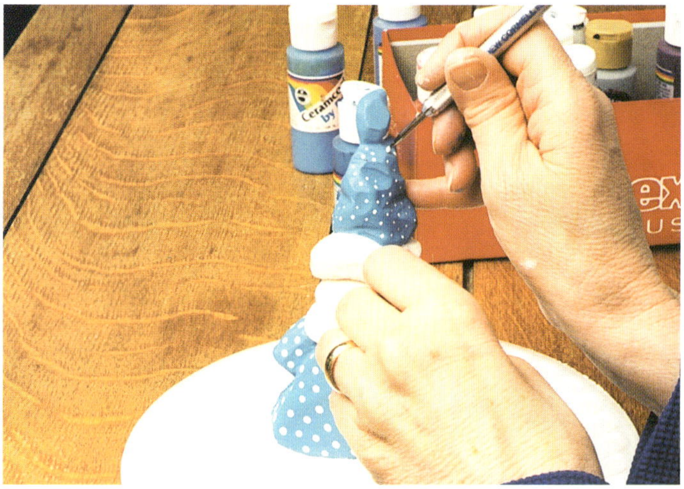

This is a smaller diameter stylus for smaller dots.

The small dots go all over the hat.

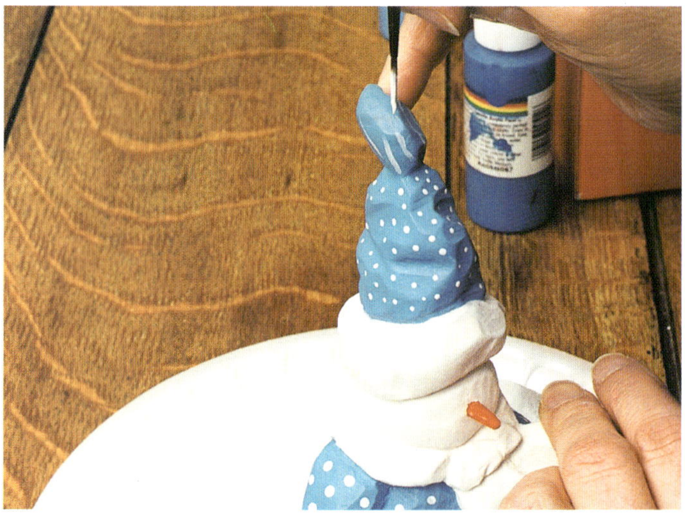

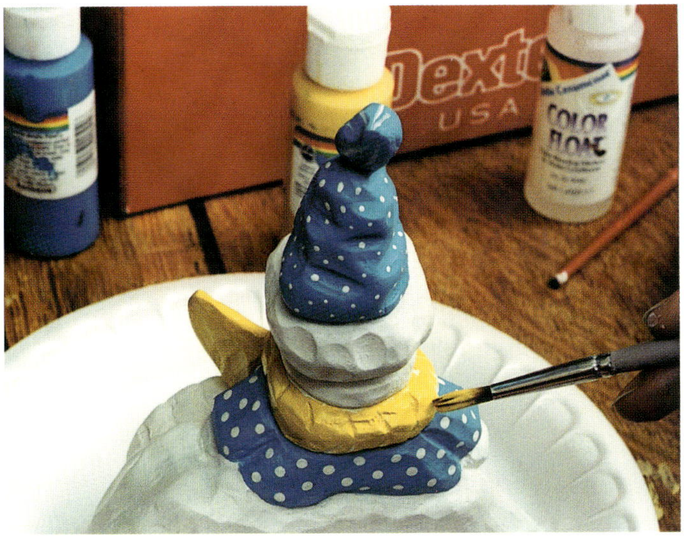

The scarf will be bright yellow.

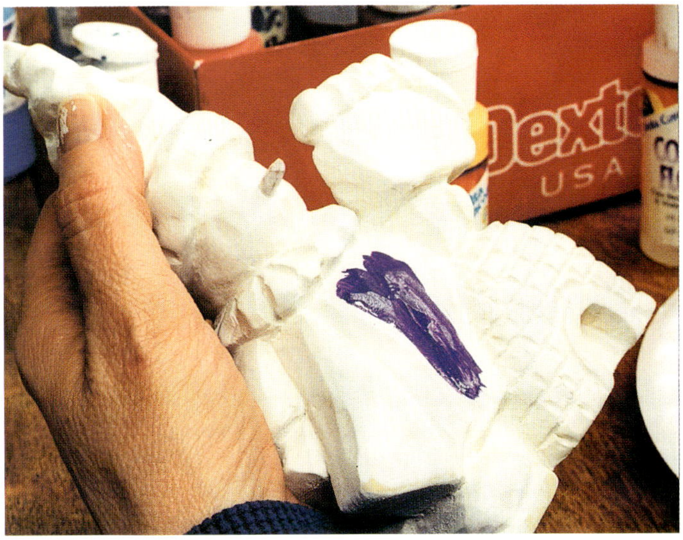

Camille shifted over to the Igloo Snowman while the other fellow dries. This snowman will have a purple coat.

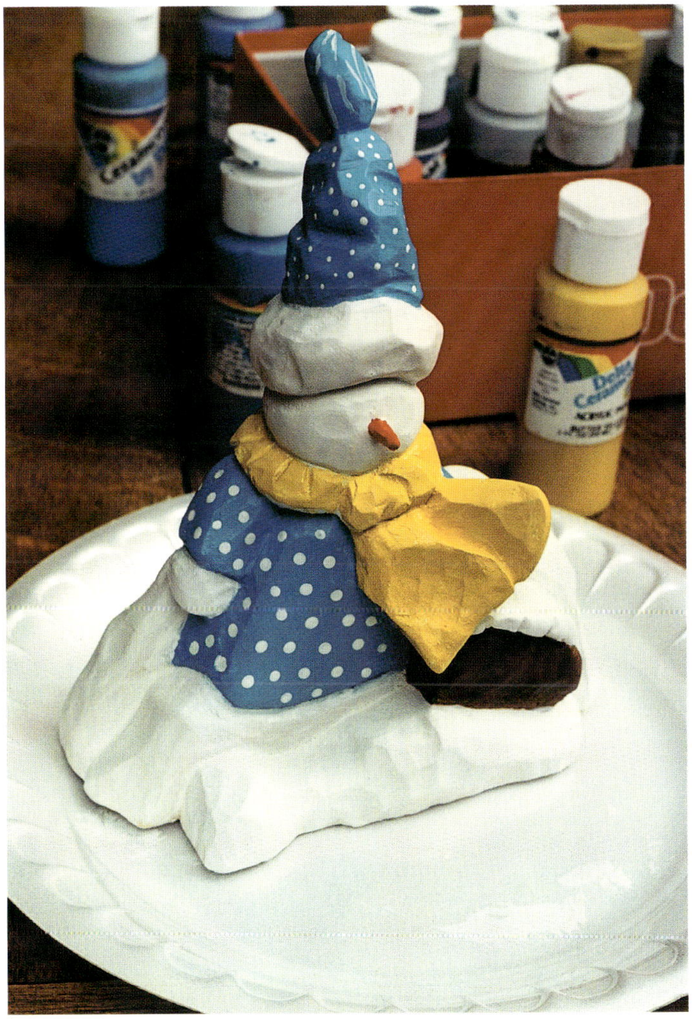

Front view of the scarf.

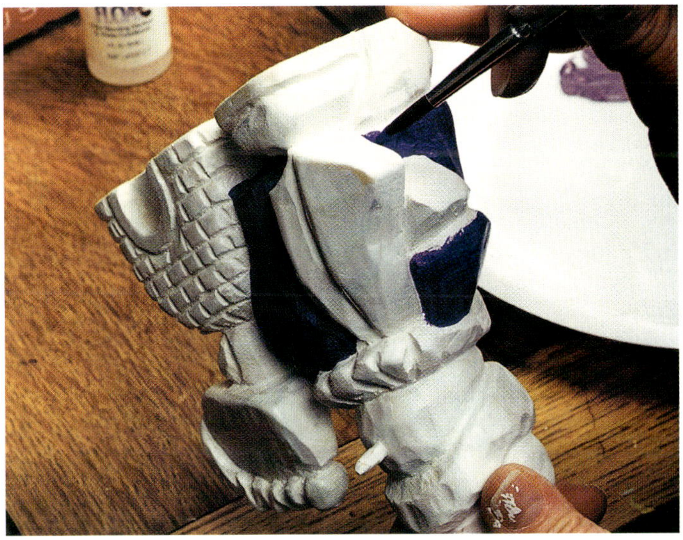

Camille paints the large areas quickly then cuts in carefully afterward.

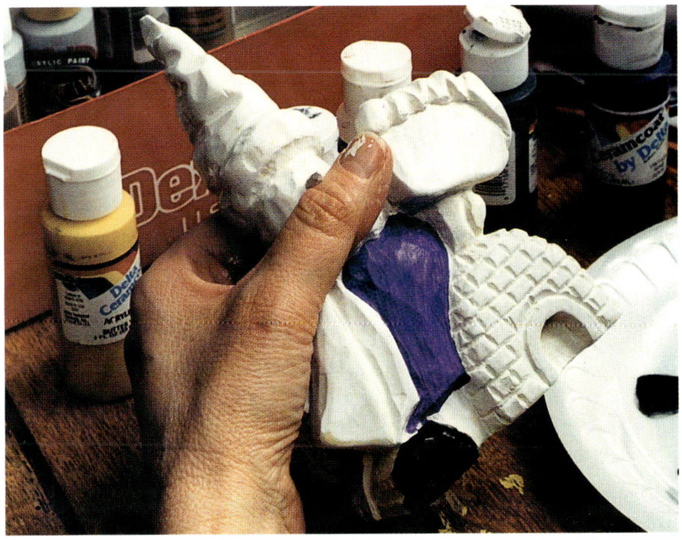

Here you see one boot has been painted black and the purple is drying.

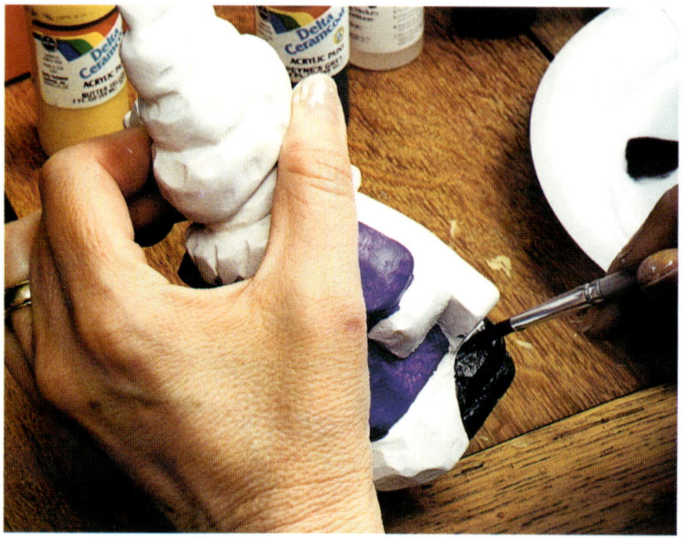

Camille finishes the boot.

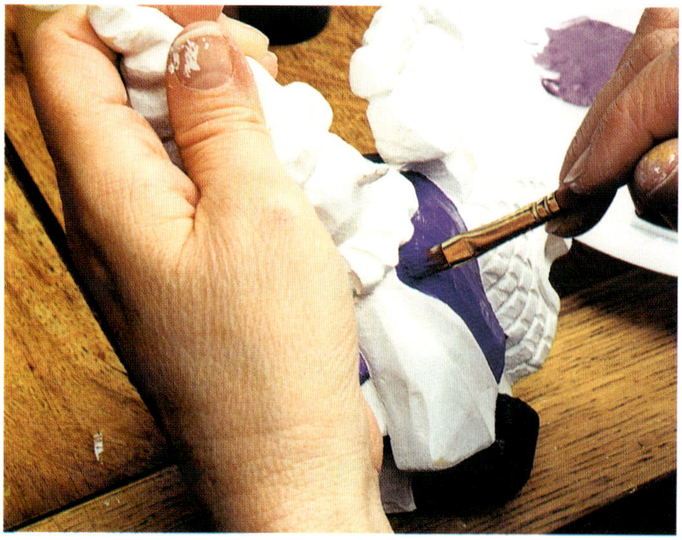

A second coating of purple goes on to the coat.

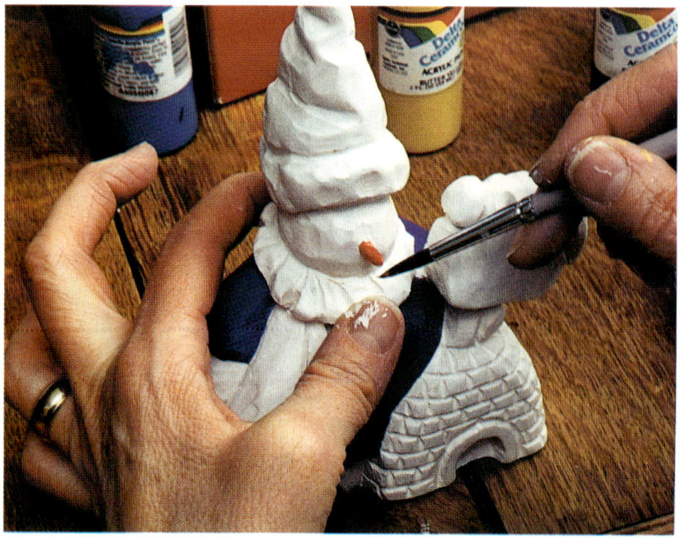

This guy gets a carrot nose too.

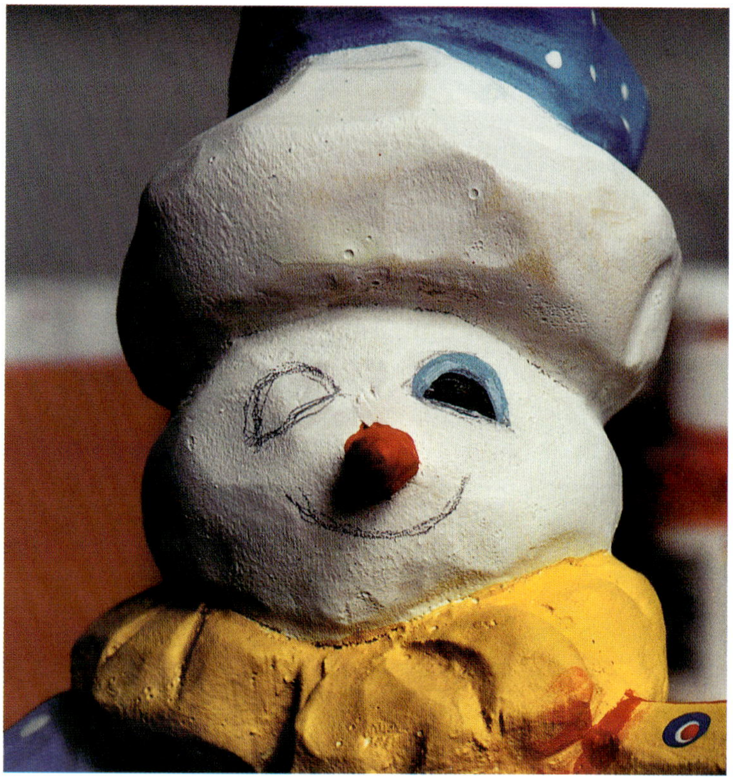

While the purple is drying on the Igloo Snowman, Camille has shifted back to the Snowbank Snowman to work on the face. You can see the pencil lines drawn on the face.

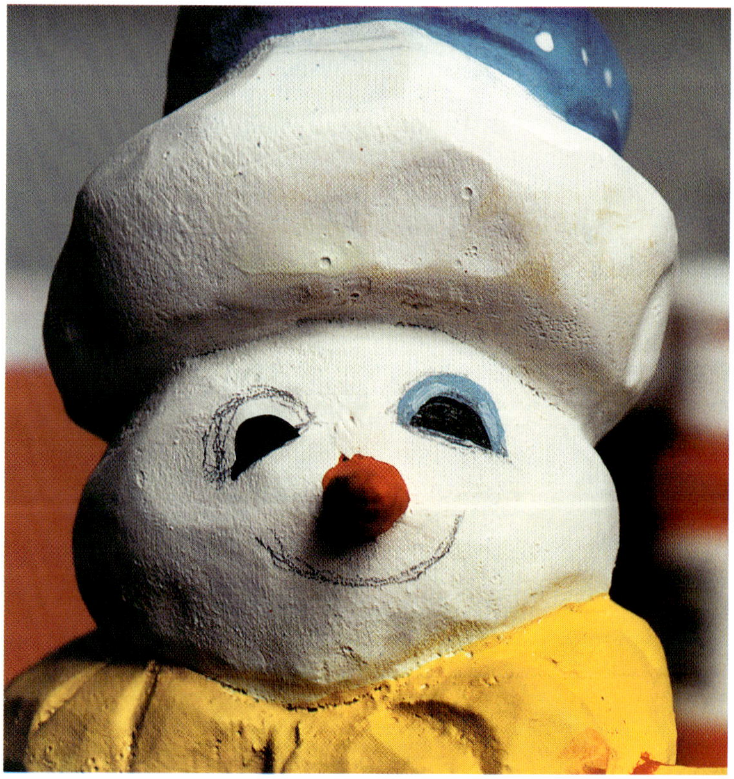

The eye starts with the black pupil.

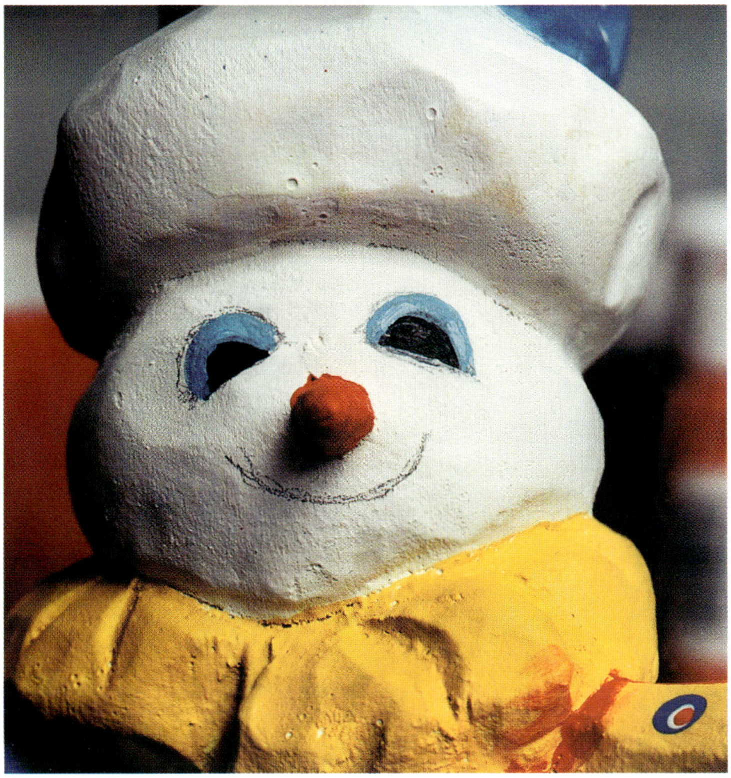

The blue goes around the black center.

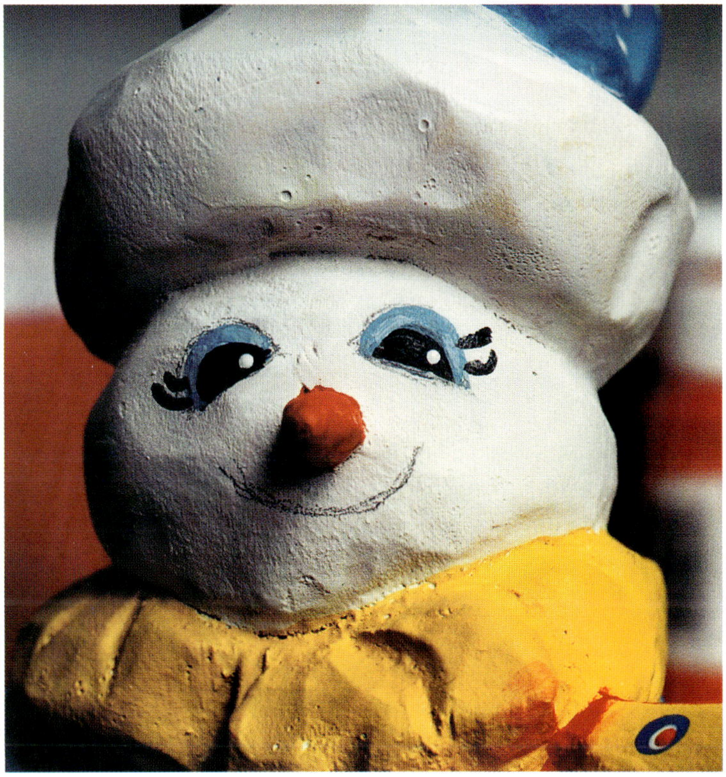

White dots are added as highlights.

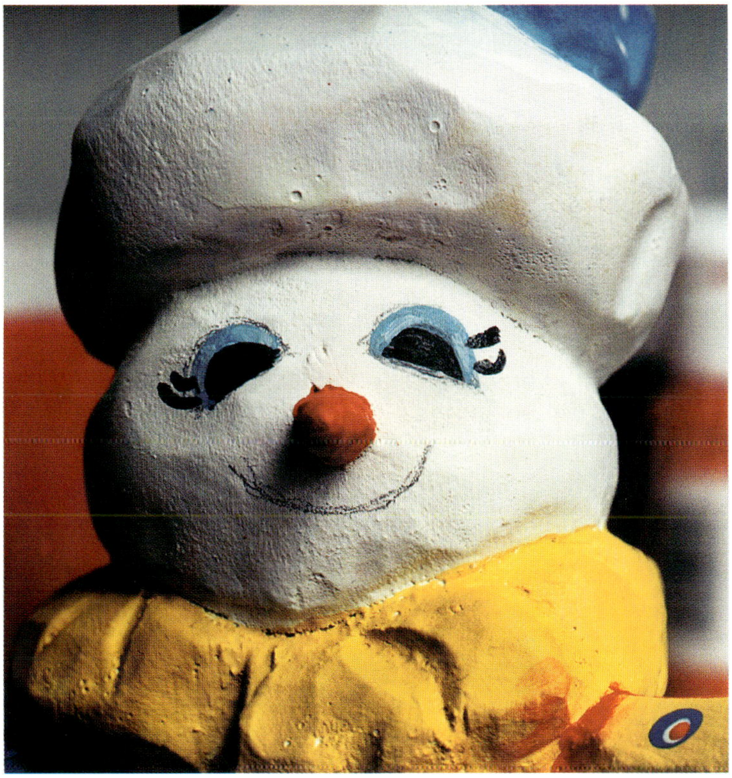

Black lines are added to the side to give the eyes cute eyelashes.

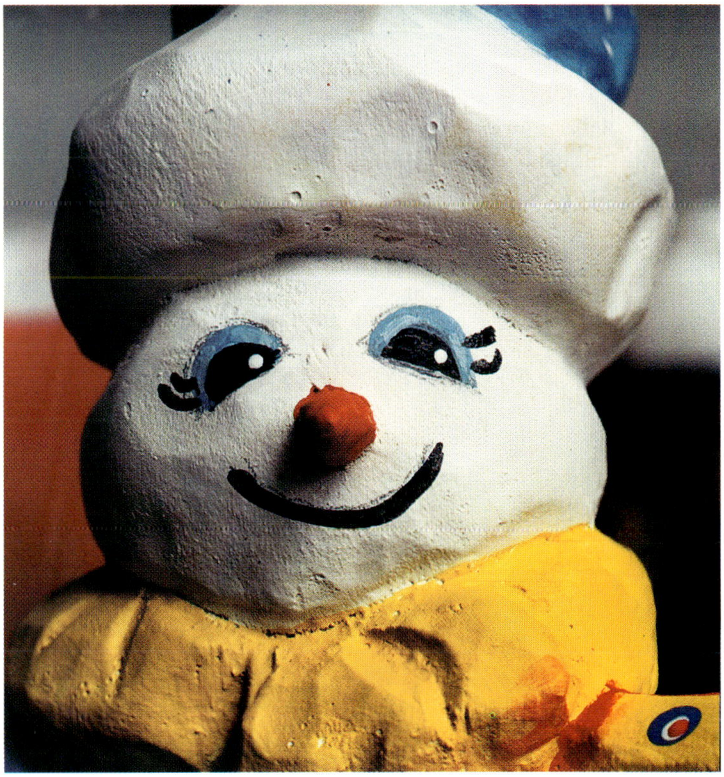

A simple one line mouth is added.

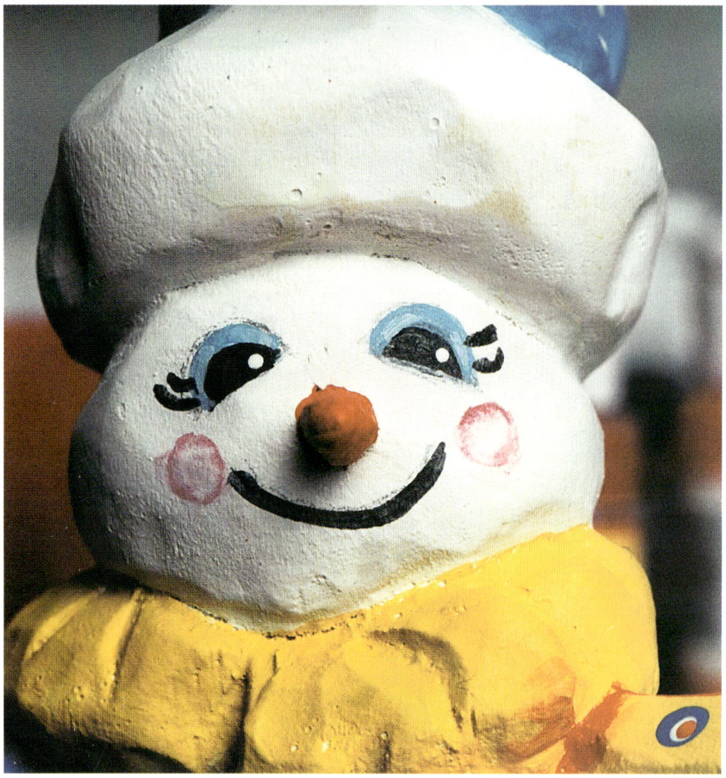

Some red cheeks are added.

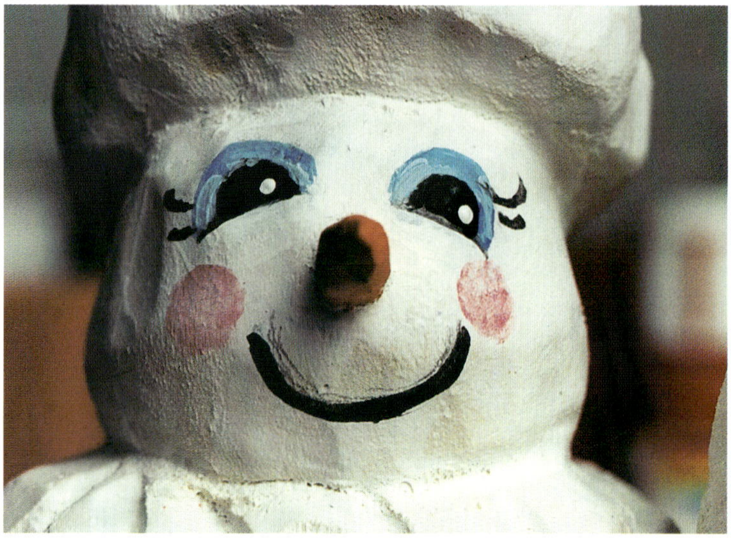

The Igloo Snowman gets eyes like the other fellow.

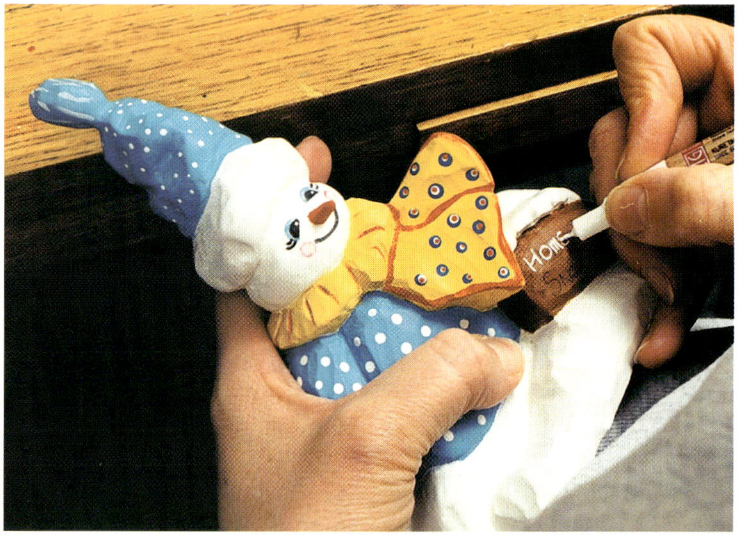

While the eyes dry, Camille letters the sign of the Snowbank Snowman.

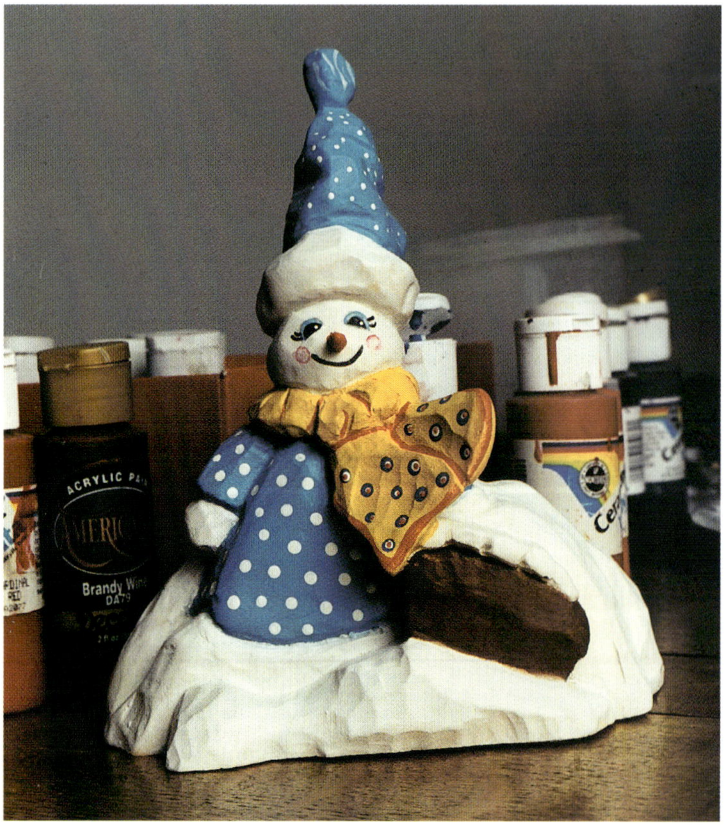

Here is the product at this point. Camille sneaked in a wild design on his scarf while we were distracted. The sign is not lettered yet but, otherwise, he is close to being finished. When we decide to put a piece into production more details are added than those shown here.

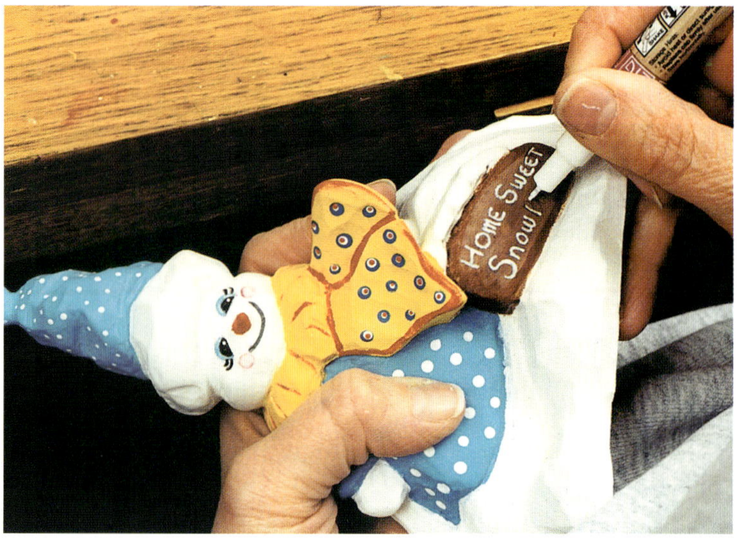

Home Sweet Snowbank.

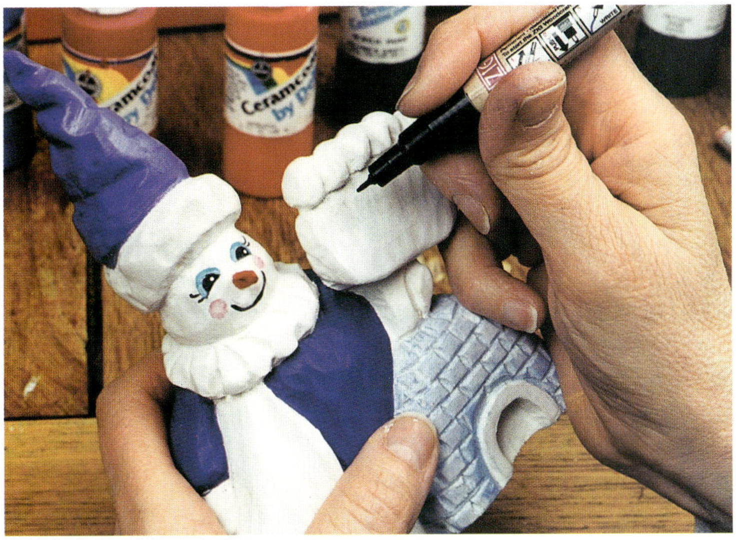

The Igloo Snowman has a sign too.

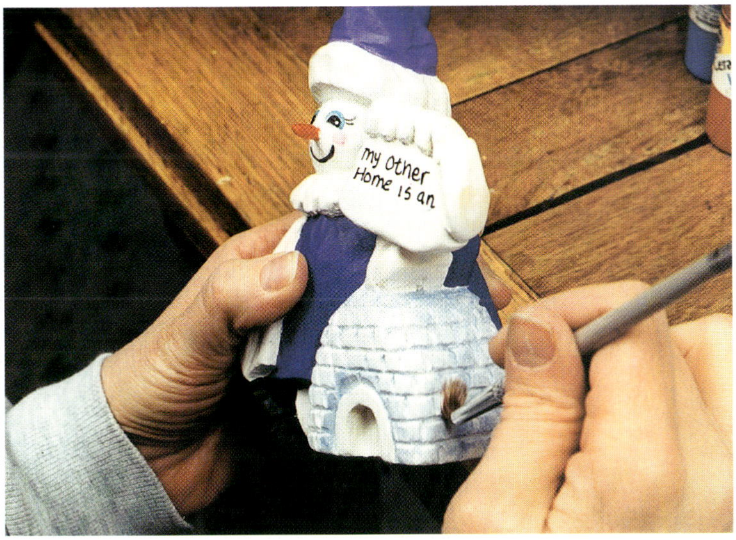

My Other Home is an... (she didn't finish the sign yet). The igloo is painted with a soft ice-blue.

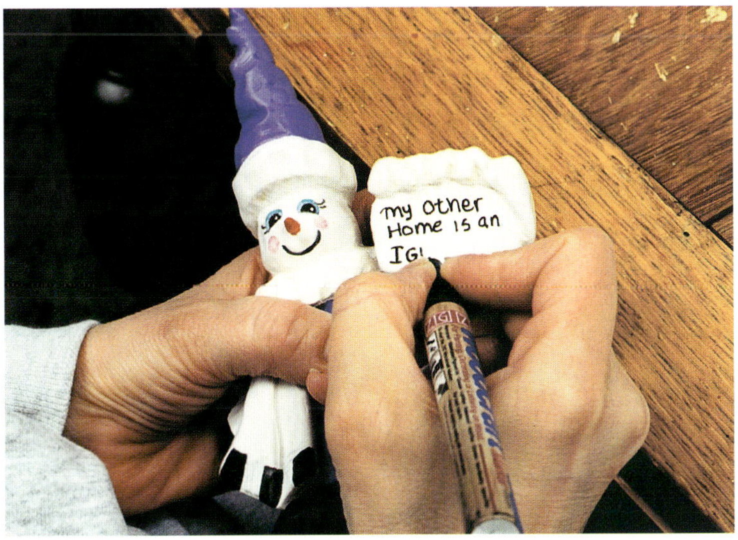

My Other Home Is an Igloo Too!

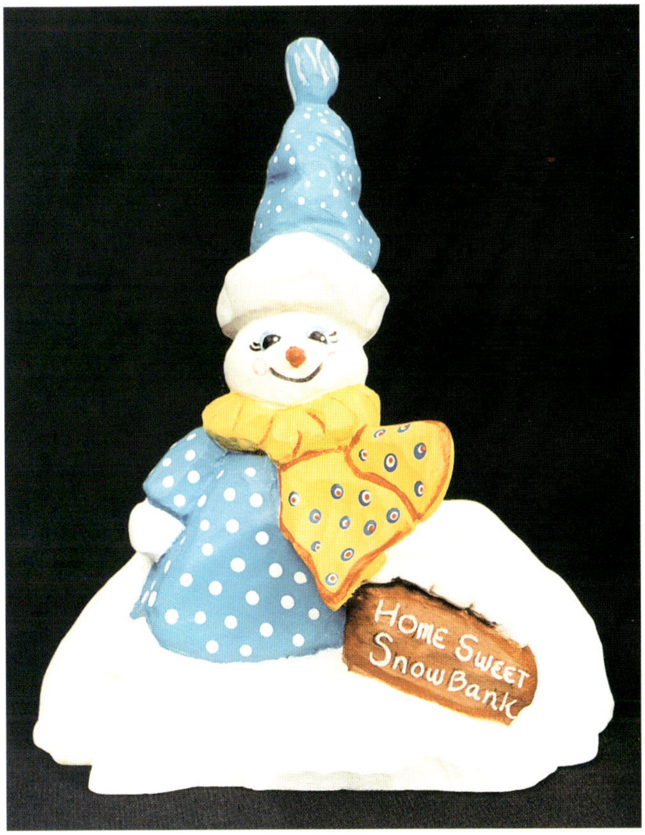

Front view of the Snowbank Snowman.

Back view.

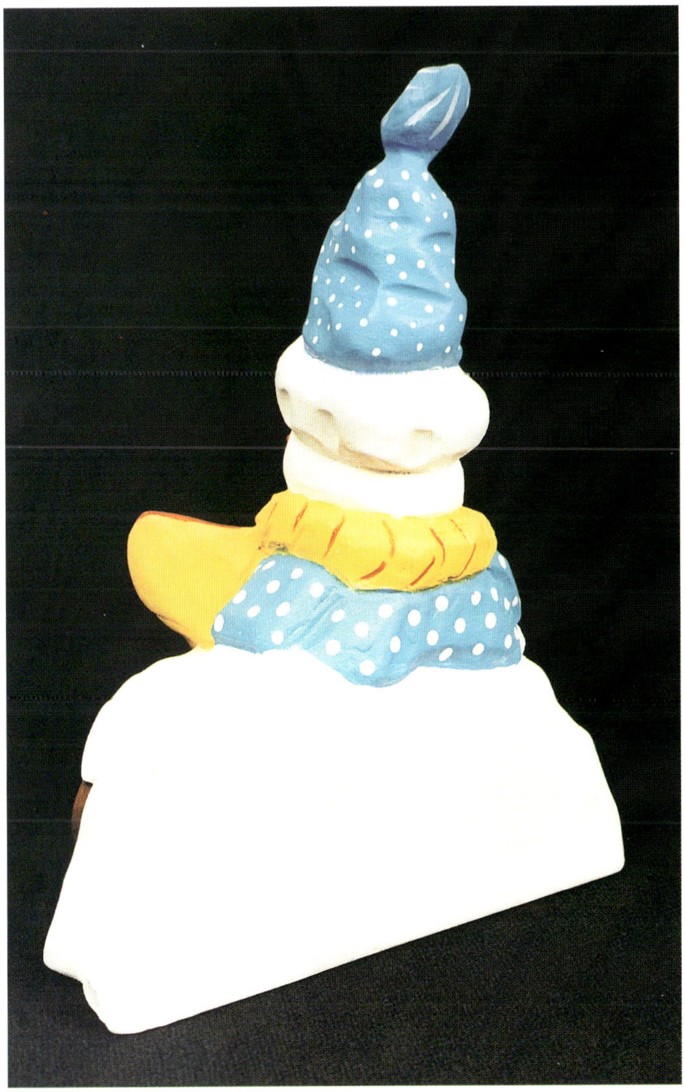

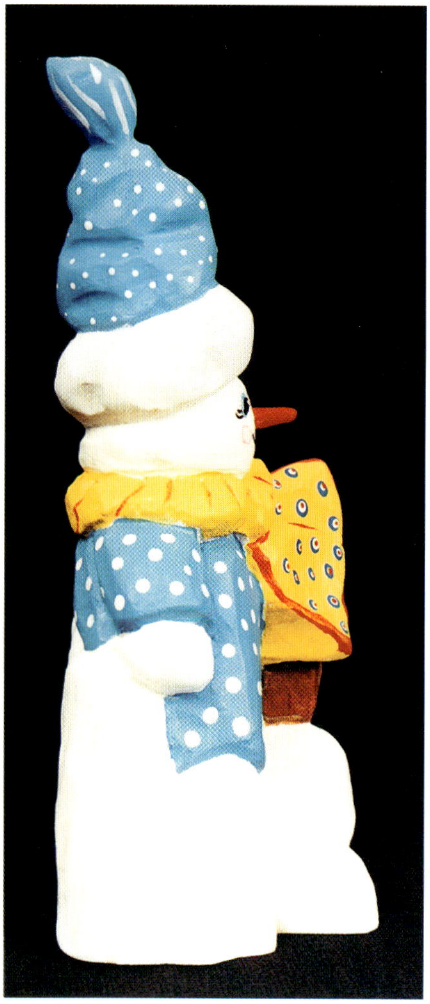

Right side.

Right:
Front view. Note that Camille put a little design at the tip of the scarf. This fellow will get much more detail before he goes into production.

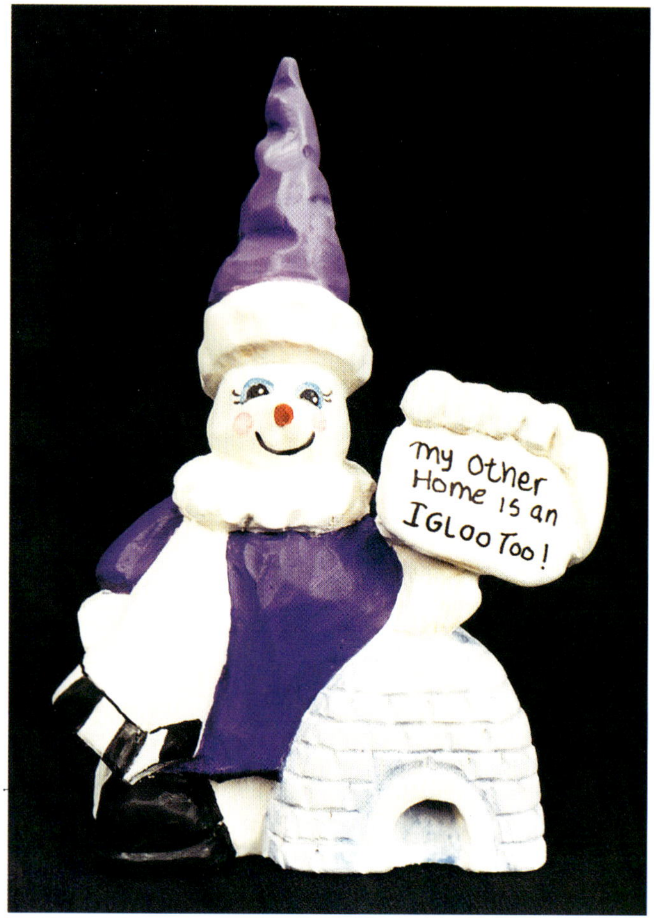

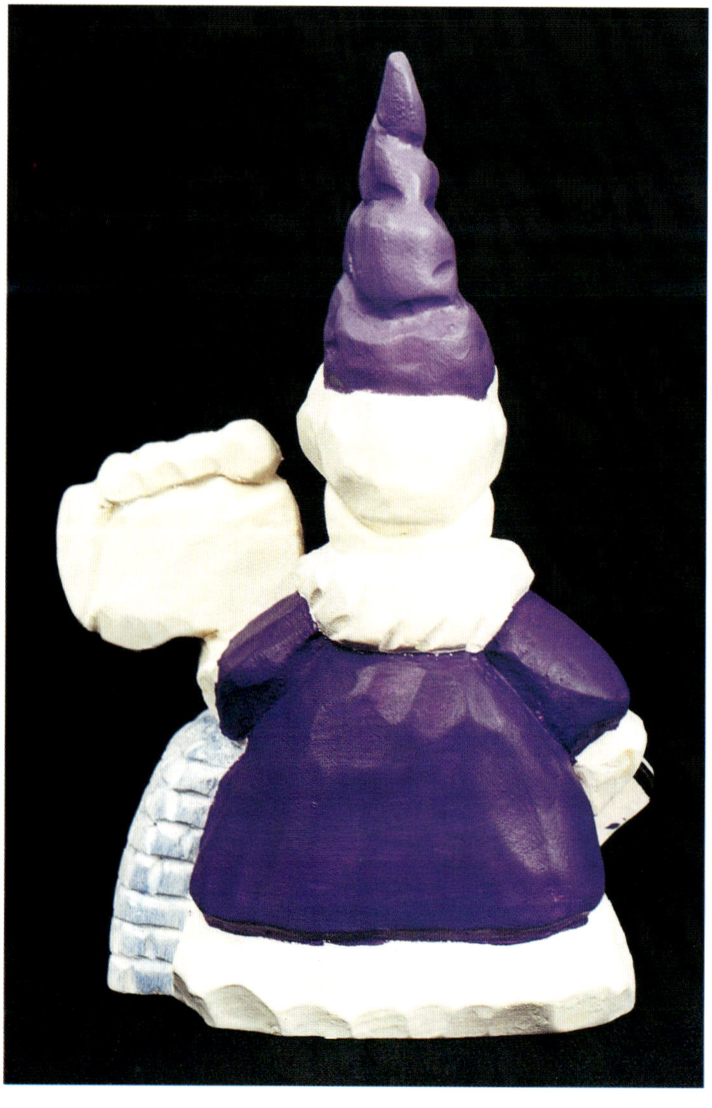

Back view.

Chapter 6

The Photo Gallery

These pictures include pieces from several different years, hence, they show different styles. This is really just a sampling of ideas that should help spark your imagination.

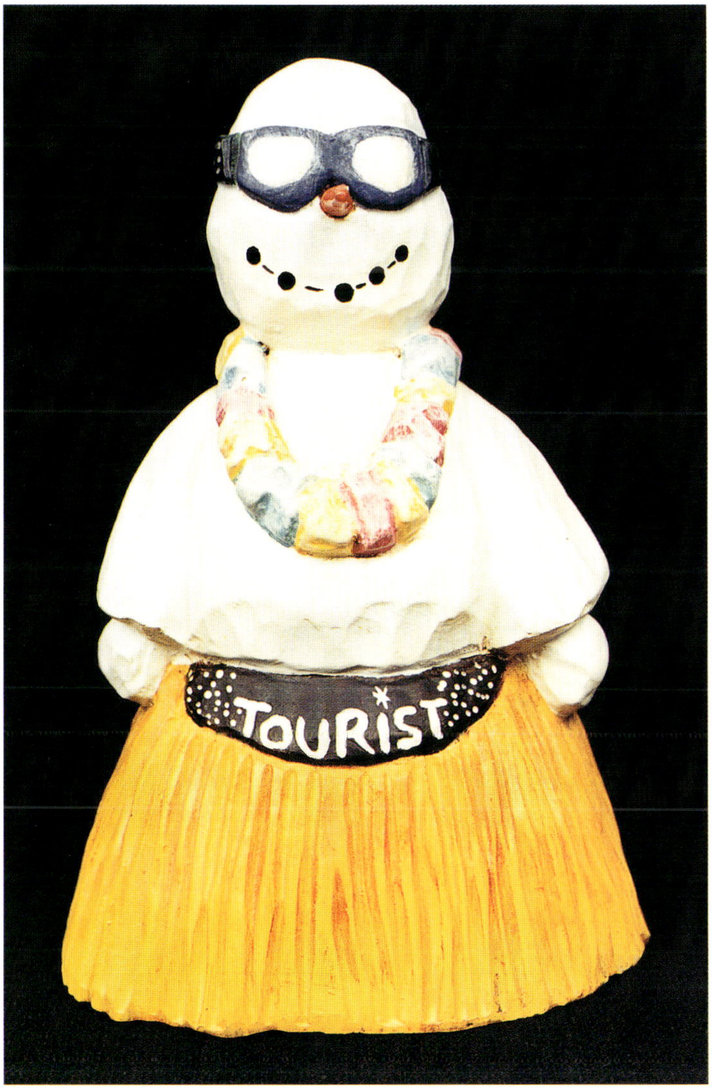

Coola Hula, the snow tourist. This snowman is decked out in a grass skirt, flower lei, and sunglasses for his trip to the islands. I wonder how long he will enjoy the warm weather?

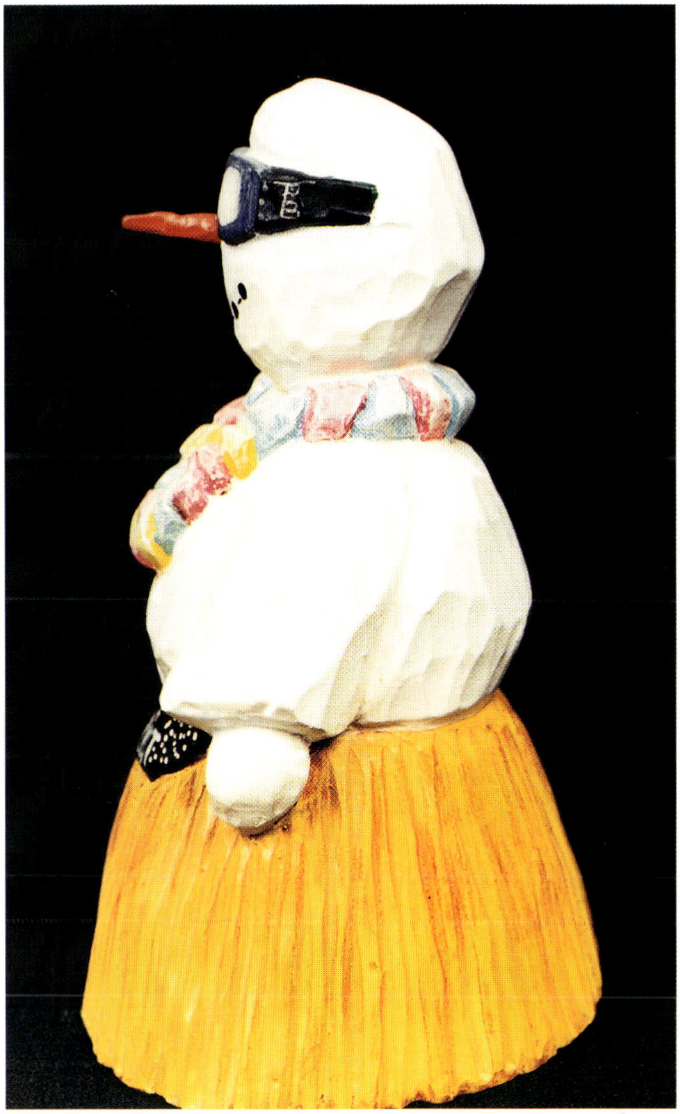

Side view of Coola Hula.

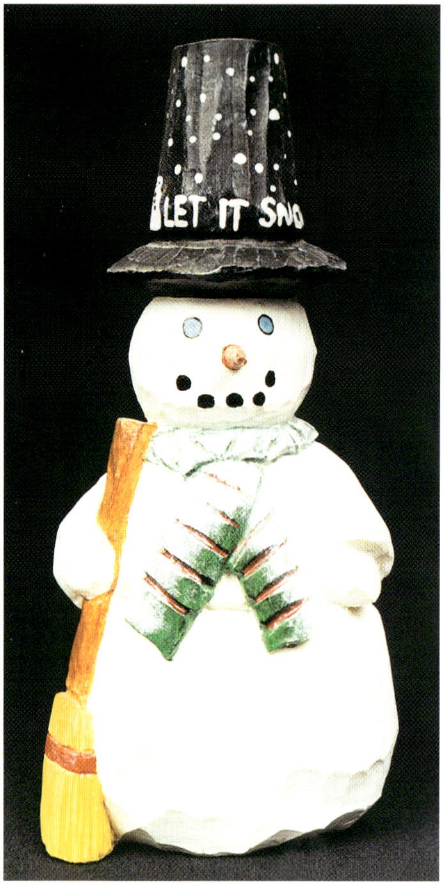

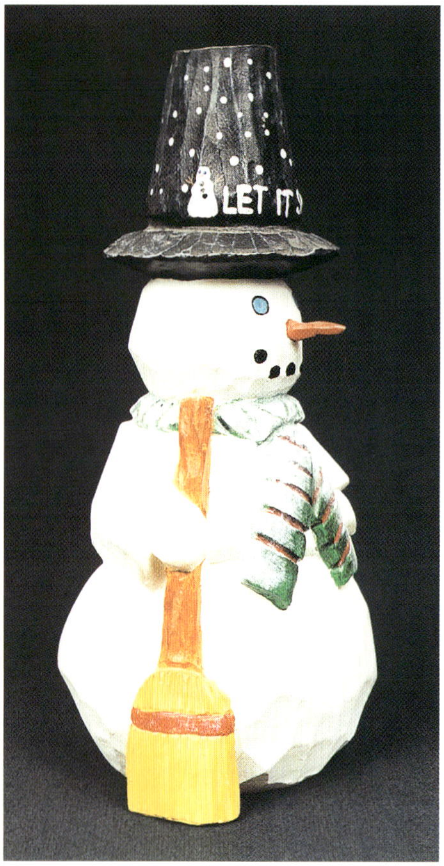

Far left:
Let It Snow. A pretty basic snowman.

Left:
Side view.

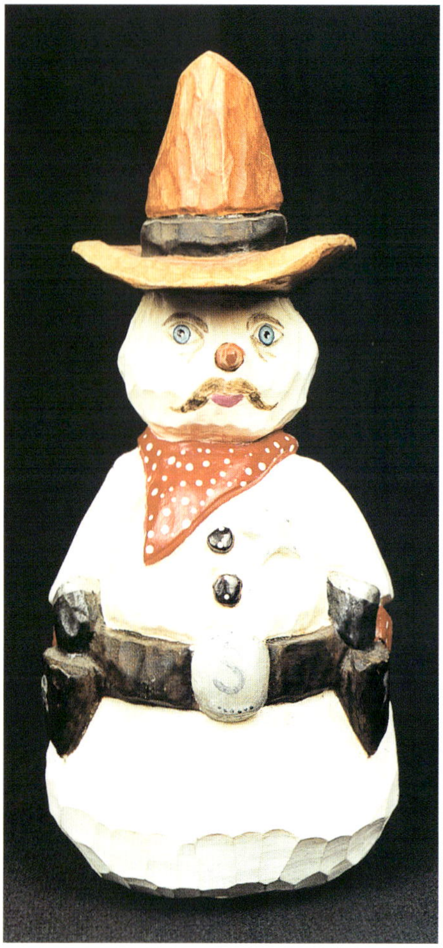

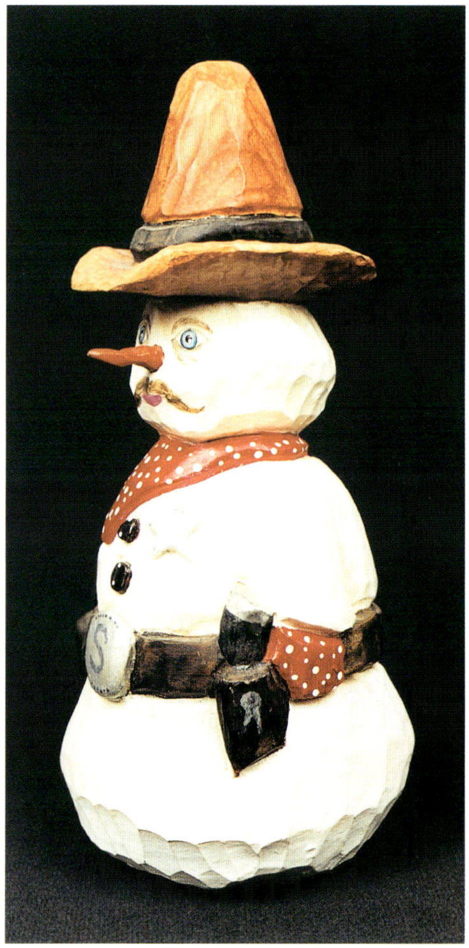

Right:
This is Snowpoke the Snow Sheriff. His job is to patrol the yard and keep the snow off the driveway. Notice the great belt buckle.

Far right:
Side view.

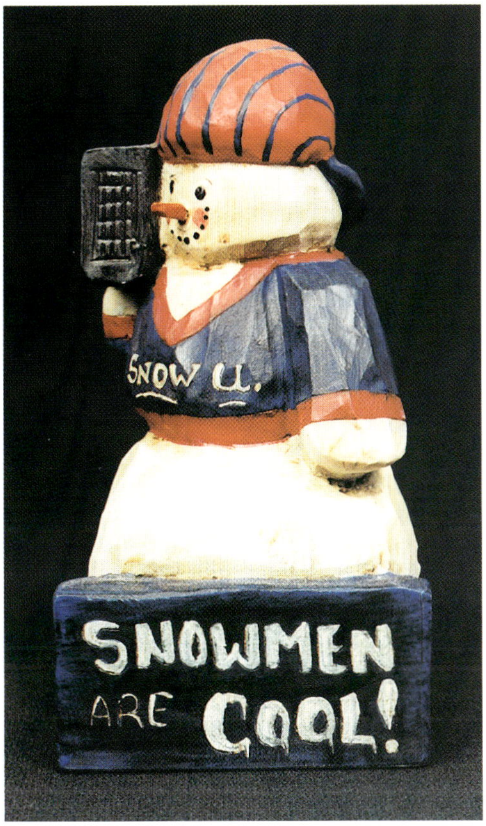

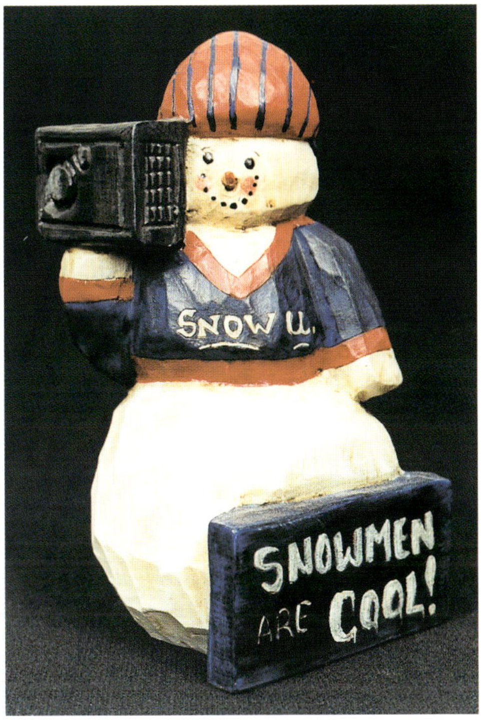

Far left:
This cool guy has his ball cap turned backwards, boombox on his shoulder, and if you look at his sweatshirt you will see that he goes to "Snow U."

Left:
The other side.

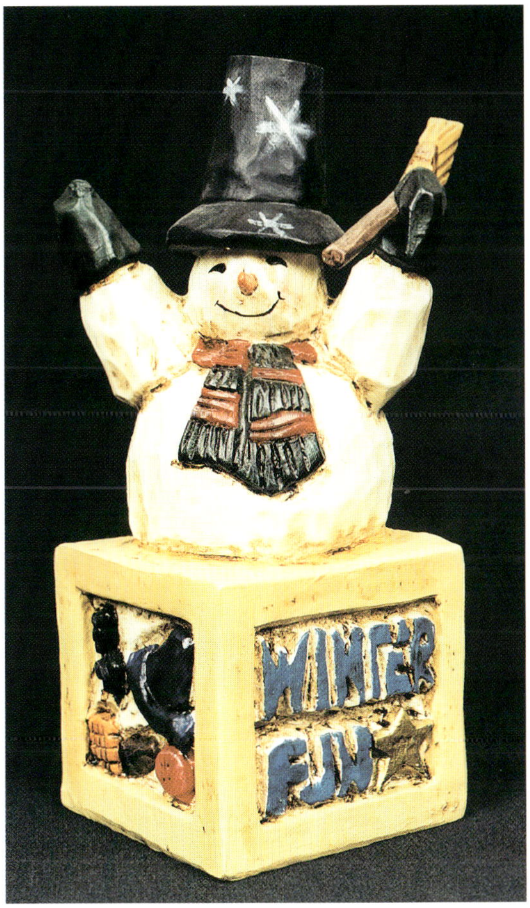

Winter Fun snowman is popping out of a box. The box has various decorations on it like snowman gear – buttons, coal, corncob pipe, etc. He is holding a broom in his hand.

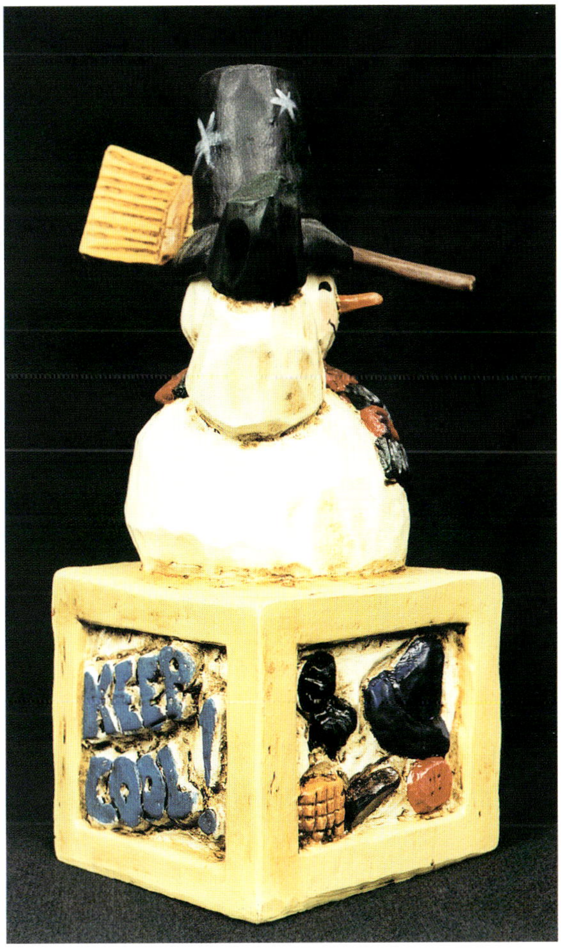

Side view.

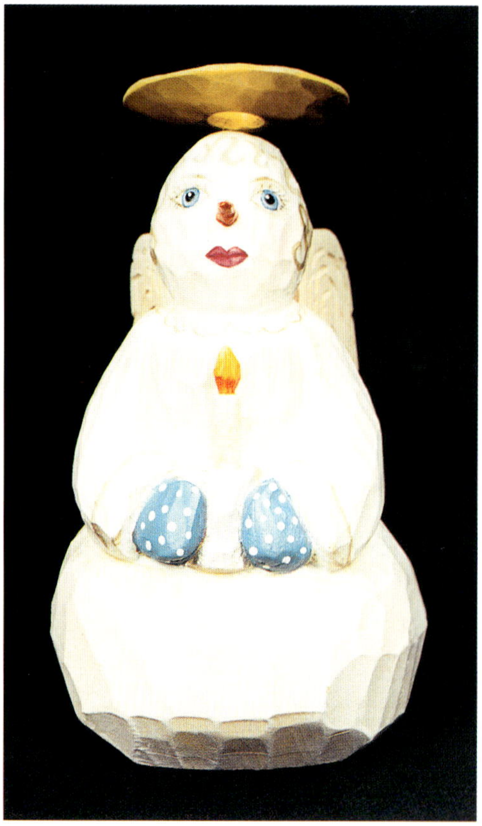

A snow angel.

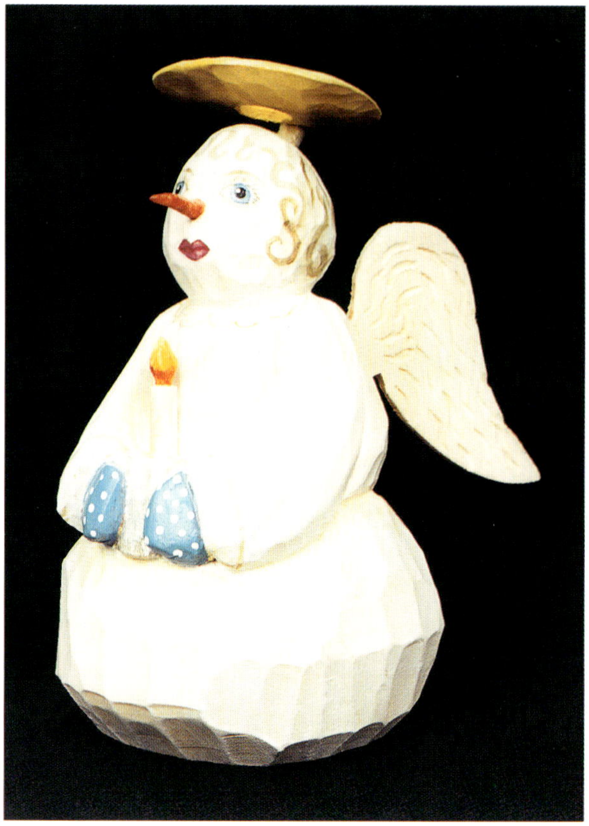

Side view.

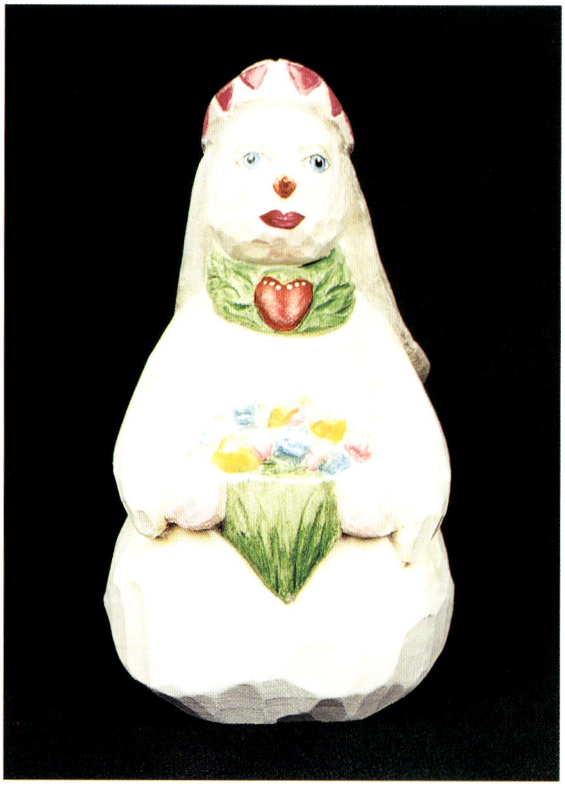

A snow bride.

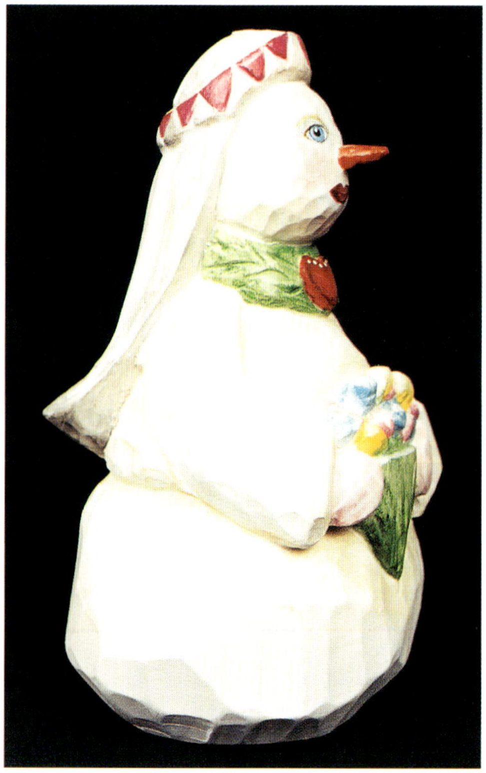

Side view.